主编_ 孔新民

中国室内设计年鉴
China Interior Design Annual

厦门特辑
2010 Xiamen Interior Design Annual

2010

⊕ 陈方晓_Frans Chan；⊕ 方国溪_Fang Guoxi；⊕ 方令加_Fang Lingjia；⊕ 胡若愚_Hu Ruoyu；
⊕ 黄少雄_Huang Shaoxiong；⊕ 黄振耀_Huang Zhenyao；⊕ 赖云舟_Lai Yun Zhou；⊕ 李泷_Leo；
⊕ 李学文_Awen；⊕ 刘腾华_Liu Tenghua；⊕ 邵力中_Shao Lizhong×杨琳_Yang Lin×姜辉_Jiang Hui；
⊕ 汤建松_Tang Jiansong；⊕ 吴伟宏_Wu Weihong；⊕ 徐福民_Xu Fumin；
⊕ 曾冠伟_Zeng Guanwei；⊕ 郑传露_Zheng Chuanlu×朱鹭欣_Zhu luxin×庄钰琳_Zhuang Yulin；

中国林业出版社

我眼中的厦门室内设计界

海上花园鼓浪屿以她自己的调子，以她自认为优雅的姿势闪耀在环东海域。清末来自欧美的列强们给鼓浪屿带来枪炮的同时也带来建筑和审美意识形态。几十栋精美典雅风格各异的别墅，在作为租界的鼓浪屿生长出来并与之融为一体。因此，鼓浪屿多了一个"万国建筑博物馆"的美名。

温润丰满的海风挟带舶来文化滋润着鼓浪屿、滋润着厦门，也滋润着厦门室内设计界。

厦门室内设计界历来都是按自己的格调不紧不慢的走着自己的节奏。"正是鼓浪屿花朝月夕，才熏陶出一颗玲珑剔透的心"。

每年几次的台风如秋风扫落叶，来了去、去了来，没给厦门带来什么也没从厦门带走什么，只让厦门一次次展现有容乃大的胸怀。而2010年由厦门设计室内界挂起的"台风"席卷着中国室内设计界，并迅速上升成为媒体所谓的"厦门现象"。此台风绝非空穴来风，它注定会留下点什么，值得分析，值得记录。

也许，我们有必要回忆来自20世纪80年代的另一场台风。中华文明的文脉从1949年移植到台湾后一直没有断裂过，甚至比大陆更是完整。这样的底蕴造就台湾成为"最中国"的室内设计流派，在某种意义上讲，厦门设计界首先受台湾设计的影响而启蒙。共同的语言、共同的气候和生活方式，在80年代国门伊始洞开时，台湾的生意人选择了在厦门投资，具有新的审美形态和新型建材工艺的写字楼、卖场、娱乐场、酒店等如雨后春笋。此时，闯入厦门的台湾设计，和来自广东的香港设计对内地产生了巨大的影响。迅速占领内地的设计、装饰市场，并在内地留下了许多不甚成熟的台式、港式风格的作品。

舶来文化、殖民地风格建筑的耳濡目染，加上此番"台风"的洗礼，加上积淀了厦门室内设计界的审美班底——西形东韵，同时培养出厦门第一代室内设计师。此时，国内还没有室内设计专业。第一代室内设计师多为美术出身。做台湾设计师的助理，帮台湾设计师描图，从实战出发，从细节开始，设计、工程两手开弓。第一代室内设计师是全才，不单精于设计，还精通工程管理造价控制，同时还有一手唯美精致的手绘

主编_ 孔新民

中国室内设计年鉴
China Interior Design Annual

厦门特辑
2010 Xiamen Interior Design Annual

2010

陈方晓_Frans Chan；方国溪_Fang Guoxi；方令加_Fang Lingjia；胡若愚_Hu Ruoyu；
黄少雄_Huang Shaoxiong；黄振耀_Huang Zhenyao；赖云舟_Lai Yun Zhou；李泷_Leo；
李学文_Awen；刘腾华_Liu Tenghua；邵力中_Shao Lizhong×杨琳_Yang Lin×姜辉_Jiang Hui；
汤建松_Tang Jiansong；吴伟宏_Wu Weihong；徐福民_Xu Fumin；
曾冠伟_Zeng Guanwei；郑传露_Zheng Chuanlu×朱鹭欣_Zhu luxin×庄钰璘_Zhuang Yulin

中国林业出版社

我眼中的厦门室内设计界

海上花园鼓浪屿以她自己的调子、以她自认为优雅的姿势闪耀在环东海域。清末来自欧美的列强们给鼓浪屿带来枪炮的同时也带来建筑和审美意识形态。几十栋精美典雅风格各异的别墅，在作为租界的鼓浪屿生长出来并与之融为一体。因此，鼓浪屿多了一个"万国建筑博物馆"的美名。

温润丰满的海风挟带舶来文化滋润着鼓浪屿、滋润着厦门，也滋润着厦门室内设计界。

厦门室内设计界历来都是按自己的格调不紧不慢的走着自己的节奏。"正是鼓浪屿花朝月夕，才熏陶出一颗玲珑剔透的心"。

每年几次的台风如秋风扫落叶，来了去、去了来，没给厦门带来什么也没从厦门带走什么，只让厦门一次次展现有容乃大的胸怀。而2010年由厦门设计室内界挂起的"台风"席卷着中国室内设计界，并迅速上升成为媒体所谓的"厦门现象"。此台风绝非空穴来风，它注定会留下点什么，值得分析，值得记录。

也许，我们有必要回忆来自20世纪80年代的另一场台风。中华文明的文脉从1949年移植到台湾后一直没有断裂过，甚至比大陆更是完整。这样的底蕴造就台湾成为"最中国"的室内设计流派，在某种意义上讲，厦门设计界首先受台湾设计的影响而启蒙。共同的语言、共同的气候和生活方式，在80年代国门伊始洞开时，台湾的生意人选择了在厦门投资，具有新的审美形态和新型建材工艺的写字楼、卖场、娱乐场、酒店等如雨后春笋。此时，闯入厦门的台湾设计，和来自广东的香港设计对内地产生了巨大的影响。迅速占领内地的设计、装饰市场，并在内地留下了许多不甚成熟的台式、港式风格的作品。

舶来文化、殖民地风格建筑的耳濡目染，加上此番"台风"的洗礼，加上积淀了厦门室内设计界的审美班底——西形东韵，同时培养出厦门第一代室内设计师。此时，国内还没有室内设计专业。第一代室内设计师多为美术出身。做台湾设计师的助理，帮台湾设计师描图，从实战出发，从细节开始，设计、工程两手开弓。第一代室内设计师是全才，不单精于设计，还精通工程管理造价控制，同时还有一手唯美精致的手绘功底。此时设计师还未真正意义上的收取设计费，设计费涵盖在工程造价中，第一代的设计师靠工程赚钱。所以现在还留在设计界的已为数不多。硕果仅存的第一代室内设计师都对室内设计专业有着强烈执着的热爱。此阶段厦门设计师的工作方式和作品都还活在台湾室内设计的阴影里。

与第一代设计师不同，科班出身的第二代设计师，成为现在厦门设计界的主力军。理论与实践知识同样丰富。让第二代设计师多了些学院味。他们有着知识份子的偏执，肩负着太多的风格和主义，步履蹒跚但是坚定，他们行走在世界各地，用眼睛度量完散布在世界各地的大师作品，他们试图用20年的时间完成了西方国家200年的室内史的历程。他们与港、澳、台一线设计师搭档完成了许多重点项目，他们还和欧美国家的一线设计事务所亲密沟通，试图拉近厦门设计界与国际化的距离。他们还肩负着培养厦门室内设计界的任务，第二代室内设计师开设设计公司成为厦门室内设计界的"黄埔军校"。

80后横空出世，他们年轻充满朝气，在"黄埔军校"完成了进入厦门室内设计界的第一步，带着理想轻装上路，横刀立马仗剑走天涯。

福建人的勤勉、守信、团结让厦门室内设计界的老中青亲密组合拿下了：2008年IAI亚太双年大赛金奖、银奖；2008年IDA国际建筑景观室内设计大赛特等奖、金奖；2009年APIDA香港亚太设计大赛银奖、铜奖；2009年IAI亚太室内设计精英设计师邀请赛金奖； 2010年AIDIA亚洲设计大赛金奖、银奖；2010年APIDA香港亚太大赛金奖、银奖、铜奖；2010年亚太双年大奖……

厦门室内设计界值得期待，中国设计界值得期待。

陈方晓

2011年3月

陈方晓 \ Frans Chan 08

<< 厦门作为沿海经济特区在20世纪80年代曾经领先过国内设计装修领域。之后陷入了漫长的停滞期，今天很可喜的看到在厦门涌现出一批优秀的设计师，其中还有设计师多次获得国际设计大奖，我们用作品赢得了同行的尊重。

- 解码DNA/Decoding of DNA 10
- 隐/Hidden 16
- 拈花一笑/Defloration Smile 22
- 星空/Stars World 30

方国溪 \ Fang Guoxi 36

<< 厦门是温馨的海滨城市，厦门城市不大，生活便利。生活、工作、压力也不那么大，所以厦门的设计有一种自然的小资"泡茶厝前、听涛看海"。中国设计近年发展很快，国内有很多设计师崭露头角，如何看中国设计？那得从国外看回来，那样看的话，中国设计又太"设计"了，我想是跟我们的思想观念有关，中国设计要在世界设计之林有一席之地的话，那得有值得人家喜欢的生活方式，说到底是文化，有值得传继的文化。

- 厦门御榕庄男子养生会所/Men's Health Club Royal Banyan Tree 38
- 厦门五缘湾样板房/Wuyuan Bay Open Houses 44

方令加 \ Fang Lingjia 50

<< 也许80年代的人都有这样的梦想，有一所房子，不一定面朝大海，但是至少也在山野之间，回归乡村和自然。我是在农村长大的，在城市里觉得拥挤、压抑，比较渴望回到农村。

- 一尊皇牛/Yizun Huangniu Restaurant 52
- 清汤餐厅/Qingtang Restaurant 58

胡若愚 \ Hu Ruoyu 64

<< 厦门之前业界的联系比较少，现在慢慢多了。相互之间的交流，可以促进相互学习。我很喜欢和年轻的设计师在一起，从他们身上我也学到了很多。

- 厦门五缘湾样板间/Wuyuan Gulf Show Flat 66
- 成都金沙鹭岛售楼处/Egret Island Sands sales offices 74

黄少雄 \ Huang Shaoxiong 80

<< 厦门设计也是中国设计的一个缩影，设计任务很多，大家都很忙，有很大的发挥锻炼空间，我们都想向国外的领先设计企业看齐，在概念设计、细部设计、标准制图、物料选配方面都要努力和提高，差距会越来越小的。

- 国贸蓝海F型样板房/Blue Ocean F-Trade Show Flat 82
- 同3组办公楼/Office Building With Tong3 88

黄振耀 \ Huang Zhenyao 92

<< 厦门是很幸福的城市，厦门有很多很出色的设计师，也有好多朋友都很有个性在做原创方向的研究，希望业主也能有这方面的思维和支持，那大家就有可能看到厦门的魅力了。中国需要"设计"，需要建立在我们自己人文基础上的设计，设计方面我们的业主和设计师都不必模仿任何国家，研究一下我们五千年的文化历程用当代的思维表现出来，就足以成为世界设计的方向标了。

- PARK.酒店/PARK.Hotel 94
- 中骏天峰会所/Zhongjun Tianfeng Club 102
- 丽湾优家售楼处/Liwan Youjia Sales offices 108

赖云舟 / Lai Yun Zhou 112

<< 保持激情是不懈的动力，我觉得除了自己对于设计的热爱，还有就是设计本身就充满创意、要求不断的求新求变，我对不同的空间，自然就会有一种新的激情，这种激情会一直激励我产生新的创意。当然了在设计中还必须考虑到业主的需求，如果脱离了生活，脱离了业主的需求，再好的创意都是无意义的，设计师就是艺术性、专业性的完成业主的品味空间。

- 惠安聚龙小镇售楼处/Sales Offices Of Hui'an Julong Town 114
- 厦门大韵天成设计公司办公室/Dayun Tiancheng Design Co., Ltd. 120

李泷 \ Leo 128

<< 厦门室内设计在国内整体的设计环境中还处于初级水准，无论是政府主管部门的大环境或从业者自身的专业素质以及对于行业本质的思考都有待提升。放到世界的大背景中，中国的室内设计存在同样的问题。

- 厦门泛华盛世/Xiamen Fanhua Shengshi 130
- 观音山商务中心/Guanyin Mountain Business Center 136

李学文 \ Awen 142

<< 如果可以我希望我与我们的设计团队能影响着厦门设计市场的发展，这是我们的目标。坚持走精品路线，来提高我们的品质。因为我觉得现在厦门设计师对设计与生活有许多不一样的看法，整体还是处于初、中级发展阶段，是一种起伏不定的状态。在以后的发展路线上看是慢慢的向精品设计发展了。

- 厦门新东方至尊家居馆/New Oriental Extreme Home Gallery 144
- 厦门庄式家居新馆/Zhuangshi Home Gallery 148

刘腾华 \ Liu Tenghua 152

<< 门设计市场的发展归结于设计师对设计的态度，很自豪我们厦门这么多优秀的同行都在为厦门设计走向国际而不屑的坚持着，同时我们需要更多的学习、交流、沟通、探讨、分享、鼓励、喜悦……

- 香槟城/Champaign City 154
- 郭氏家居/Mr. guo Room Design 160

邵力中 \ Shao Lizhong × 杨琳 \ Yang Lin × 姜辉 \ Jiang Hui 164

<< 我们不建议将建筑的室内外分割开讨论，这应该是一个完整的创作过程。项目完成之后所呈现的，不仅仅是风格和美感，同时还有其不可或缺的社会属性。如何取得一个项目在社会、经济、文化和美学等各个范畴的综合平衡，正是我们的课题。

- 九间房七方院/Rooms & Courtyards 166
- 厦门瑞景商业广场/Commercial center of Xiamen Ruijing 176
- 石狮建明国际酒店/Jian Ming International Hotel of Shishi 182

汤建松 \ Tang Jiansong 192

<< 厦门设计行业起步比较晚，但是这几年来得到了很大的发展。厦门设计的差异比较大，风格各异，设计师之间缺乏交流。中国设计行业发展迅速，但是专注于设计的年轻设计师生存压力也比较大。没有名气，就很难接触到好的项目；为了生存，又不得不接受普通的项目。酒香也怕巷子深，所以我们现在会尽量参加比赛。

- 东方喜意有限公司/Dongfang Xiyi Ltd. 194
- 香港和丰设计公司办公室/Hong Kong Design Co., Ltd. 200

吴伟宏 \ Wu Weihong 206

<< 每当看见一个作品诞生，重视别人的评价，无论是好还是坏，都是一种欣慰。这些观点，也是别人对我们设计的一种理解。这就是一种动力，一种设计的源泉。我们每次接到一个好的或者是不好的案子，我们都认真对待，认真对待每一件事情，希望作品尽早诞生。

- BAG办公/BAG office 208
- Bellagio餐厅/Bellagio Restaurant 214
- Show吧/Show Bar 220

徐福民 \ Xu Fumin 226

<< 如果暂时不做设计，自己最想去欧洲学习，感受生活。在理论的学习上，重新学习"建筑史"的相关课程，从理论上再提高自己，努力把空间的感念做到室内设计中。

- UTOP优伯科技办公楼/Technology office UTOP 228
- 中骏·财富中心售楼处/Fortune Plaza Sales Center 232

曾冠伟 \ Zeng Guanwei 238

<< 厦门是一个有历史积淀的城市并具备包容的文化底蕴，厦门设计应该走出去，中国设计亦然。

- 厦门U空间精品酒店/U space for boutique hotel 240
- 巴厘香墅江府/Bali Villas 246

郑传露 \ Zheng Chuanlu X 朱鹭欣 \ Zhu luxin X 庄钰璘 \ Zhuang Yulin 252

<< 厦门的市场较小，并不成熟，好的设计作品不易实现。每年我自己比较满意的项目就只有一两个。厦门业界的交流不多，没有形成切磋和互相学习的氛围。我自身也很少跟业界沟通，我们公司也很少自我包装、推广。为了接触更多的项目，我们也在慢慢的改善。比如参加比赛。这也是对自己的交代，是对自己的肯定。

- 黎柏洋服/Libo clothing store 254
- 厦门共想装饰设计公司办公室/Gongxiang Decoration Design Co., Ltd. 260

我眼中的厦门室内设计界 02

厦门设计的觉醒 266

2010
厦 门 特 辑
2010 Xiamen Interior Design Annual

陈方晓 \ Frans Chan

- 陈方晓设计事务所 创作总监
- *Frans Chan Design International_Creative Director*

个人经历
- 陈方晓设计事务所 创作总监
  香港战神装饰陈设顾问有限公司 创作总监
  阿拉伯酋长国迪拜阿扎曼艺术大学 客座教授

获奖情况
- 2008亚太室内设计双年大奖赛商业空间 《星空》_金奖
  2010亚洲室内设计竞赛样板房类 《隐》_金奖
  2010亚洲室内设计竞赛样板房类 《拈花一笑》_银奖
  2010亚洲室内设计竞赛样板房类 《天书》_银奖

Experience
- Frans Chan Design International_Creative Director
  Hk Ares Design Consultant Limted_Creative Director
  Ajman University Of Science &Technology_guest Professor

Awards
- Asia Pacific Interior Design Biennial Awards 2008 《Star World》_Champion
  Asia Interior Design Award 2010 《Hide》_Golden
  Asia Interior Design Award 2010 《Flower And Smile》_Silver
  Asia Interior Design Award 2010 《A Sealed Book》_Silver

厦门作为沿海经济特区在20世纪80年代曾经领先过国内设计装修领域。之后陷入了漫长的停滞期，今天很可喜的看到在厦门涌现出一批优秀的设计师，其中还有设计师多次获得国际设计大奖，我们用作品赢得了同行的尊重。

贵司主要从事哪方面的设计？最新的作品是什么？

- CDI（陈方晓C设计师事务所）十多年一直从事地产及地产配套产品的设计研发，如建筑外观的视觉整合、建筑扩初前的户型优化、售楼会所、样板房的设计等。
- 最近的新作品是万科地产厦门公司写字楼、香山国际游艇S型会所、金都海湾置业之厦门环东海域"海尚国际"户型优化项目、中骏置业之中骏"天峰"项目等。

您的设计灵感来源于什么？您的设计理念是什么？

- 一花一世界，一叶一如来。一花一叶一草一木都是我设计灵感的来源。在我看来设计就如同讲故事。我的每一个设计作品都是一个故事："拈花一笑"、"涟漪"、"临风玉树"、"天书"、"隐"、"星空"、"口暮"、"如意"、"九度"……
- 用戏剧的叙事手法做室内设计，让氛营造出近乎于完美的空间景象，空间的移步换景、虚实对比，空间的借和换，再加上形体、色彩、光影的渲染，就如同起伏跌宕的小说情节，让"听故事"的人深陷情节当中。

您从事设计工作多久了？

- 22年的设计生涯让我感悟良多，我见证了中国室内设计的全过程。西方国家用200年走完的路，我辈只用20年完成。虽然不可避免的有着急功近利，但是通过我辈超负荷的思考和创作，让人欣慰的是在许多领域我们已和国际同步了。

在设计过程中，您最注重哪方面的工作？对于多种设计元素有没有什么偏好？

- 我认为设计就是创意。我的作品传递了我对多元化设计元素的理解和驾驭能力。

从前期接触项目、出概念、做方案到完成整个项目，觉得最大的困难是什么？

- 一个设计项目的完成不能把交付施工图作为设计的终结。施工中项目的跟进是非常重要的，也是最难的。此时设计师更像电影导演，要能调动所有参与项目的部门协同作战，同时还要把控成本材料等等，否则不会有一个完美的作品产生。

欣赏的或者对您影响最大的人是谁？抑或某种风格、思潮、理念？

- Phillipe Starck——向Starck学习，我努力让自己作到：概念性、前瞻性、还必须有深厚的文化底蕴、丰富的人生阅历、敏锐的洞察力。还得善于从生活中汲取独特的设计元素，还必须对潮流有非凡领悟能力，引领潮流，成为潮流风向标。

怎么看待厦门设计？以及中国设计？

- 对于中国设计来讲我们这一代就是下地狱者。也许等到我们儿子这一代——90后会享受到我们这一代沉淀思考和身体力行所带来的成果。
- 厦门作为沿海经济特区在20世纪80年代曾经领先过国内设计装修领域。之后陷入了漫长的停滞期，今天很可喜的看到在厦门涌现出一批优秀的设计师，其中还有设计师多次获得国际设计大奖，我们用作品赢得了同行的尊重。

未来的公司会如何发展？您本人呢？

- 很幸运我的爱好和我的工作是同一件事，我会好好的享受设计带给我的快乐。

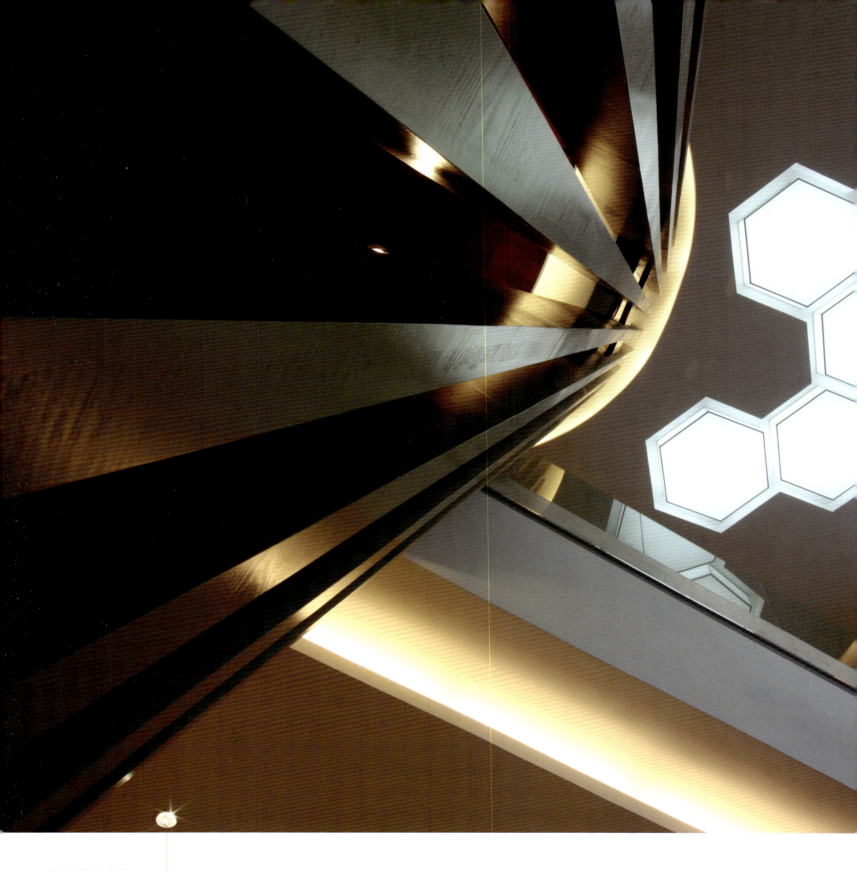

解码DNA

项目地点：福建厦门
主案设计：陈方晓
项目面积：3790 m²
竣工时间：2010年

设计概念起源于业主企业的服务内容：为各个不同企业的核心运营过程设计电脑程式，就如同DNA编码。DNA被符号化，解码为设计的主要文脉，贯穿整个空间的始终。

DNA解码为方便务实的办公操作台，一个办公组团连接另一个办公组团，将窄长的办公空间很好的化解为一个有机的整体。
DNA解码为接待大厅形象墙上的艺术装置——业主与其客户的LOGO相濡以沫。

代表作品（一）

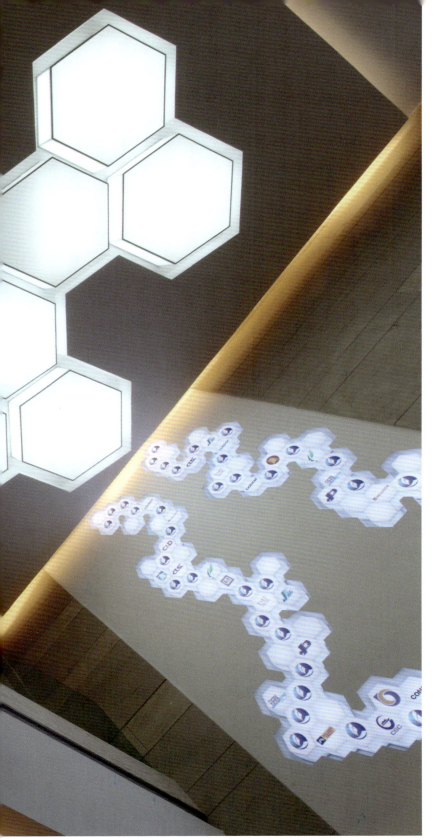

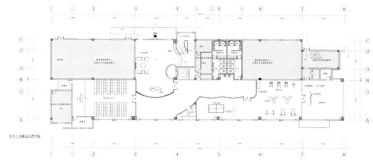

/ 一层平面图 /

/ 二层平面图 /

Decoding of DNA

Design concept originated in the owners of the enterprise services: for each different company's core operations process design computer programs, just as DNA encoding. DNA was symbolic, decoding the main context for the design throughout the entire space has always been.

DNA decoding the office desk to facilitate the practical, a group connected to another office tour office, office space will be long and narrow well resolved into an organic whole. DNA decoding the image on the wall for the reception hall of the art equipment - the owners and customers LOGO each other well.

隐

小隐隐于山，大隐隐于世；入户花园的小桥、游鱼怪石以及贴着墙缓缓而下的瀑布，都透露这"隐"的意向，勾勒出一幅闲情雅致的世外桃源；客厅的墙面"隐"藏了五扇进入厨房、卫生间、卧房的门，隐藏的门让空间整体、利索。客厅和餐厅之间纵向伸展的隔墙，透与半透之间营造出别有洞天的开阔。餐厅的圆形透窗，使人在品尝美味的同时，欣赏到墙上"隐"着的迭水口，无声无息的把水送入静止的锦鲤旁边，此时思绪也仿佛忽然凝结而顿悟……

Hidden

If you want to conceal yourself in a place, cities are better than hills. Small bridges, active fishes and the waterfalls combined together give us a sense of "concealment", just like Xanadu. The wall of a parlor conceals five doors—kitchen, washroom, bedrooms and so on, so as to make the roomage more tidy and cleaning. A partition wall between parlor and dinning room seems to widen the original space, and it looks like transparent or semi-transparent. The round window in dining room will let you enjoy the hidden water-gap when you are getting your delicious food. The water is flowing quietly till cyprinoid pool, thus your thinking may also stop to consider everything.

项目面积：161 m²
项目地址：厦门
竣工时间：2009年5月
设计师：陈方晓

Xiamen Interior Design Annual | 17

/ 平面图 /

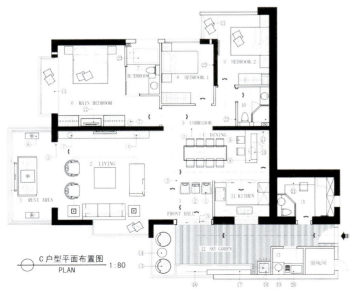

空间区域划分	
① 玄关	FRONT HALL
② 客厅	LIVING ROOM
③ 休闲区	REST AREA
④ 餐厅	DINING ROOM
⑤ 过道	AISLE
⑥ 主卧	MAIN BEDROOM
⑦ 主卫	MAIN BATHROOM
⑧ 次卧1	BEDROOM 1
⑨ 次卧2	BEDROOM 2
⑩ 次卫	SECOND BATHROOM
⑪ 厨房	KITCHEN
⑫ 空中花园	HANGING GARDEN
⑬ 洗衣房	LAUNDRY ROOM
⑭ 佣人房	SERVANT ROOM

家具指示	
① 艺术椅	ART CHAIR
② 化石木	FOSSIL WOOD
③ 花格	GRILLE
④ 艺术品	ART
⑤ 案台	TABLE
⑥ 陈设台	FURNISHINGS TABLE
⑦ 插花	IKEBANA ARRANGEMENT
⑧ 吧台	BAR
⑨ 吧椅	BAR CHAIR
⑩ 洗涤器	SINK
⑪ 酒柜	WINE CHEST
⑫ 电视柜	TV CABINETS
⑬ 软垫	PAD
⑭ 休闲茶几	LEISURE TABLE
⑮ 户外休闲椅	OUTDOOR LEISURE CHAIRS
⑯ 水景	WATER
⑰ 兵俑	BOOK
⑱ 玻璃灯箱	GLASS LIGHT BOX
⑲ 洗衣机	WASHING MACHINE
⑳ 洗衣池	LAUNDRY TRAY
㉑ 书桌	DESK
㉒ 装饰画	DECORATIVE PAINTING
㉓ 泡澡池	BATH POOL

C户型平面布置图 PLAN 1:80

拈花一笑

佛祖登座拈花示众，人天百万悉皆罔措。独有金色头陀破颜微笑。佛祖云：我有正法眼藏，涅槃妙心，实相无相，微妙法门，咐嘱摩诃迦叶。佛祖所传之乃是详和、宁静、安闲、美妙之最高心境……

佛祖拈之波罗夷花今何在？且以我之手将此种纯净无染之心境、淡然豁达之意念借无色之灰色如檀香般漫延无处不在，再引坦然自得漫于墙泄于案浸于行走之光影间，轻吟慢颂只属于东方之气韵，且气韵生动……

不能言，不可言，恍若间感悟如"涅槃"……

项目面积：168 m²
项目地址：厦门
竣工时间：2009年5月
设计师：陈方晓

Defloration Smile

Buddha took his seat with a flower in his hand and hundreds of people were confused except gold Buddhist monk. The Buddha said: "I have many rules but I just want to give Mohejiaye some words." All the beliefs that Buddha spreads are harmony, quietness, beauty and other happy things.

Where can we find the flower nowadays? I would like to do you a favor: call for quietness and inclusiveness by designing a comfortable room in order to show the oriental elegance.

No words, even not being able to say any words, which is just like "nie pan"(Buddha words, it means a perfect world for the Buddha monk)

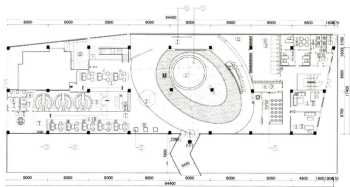

"泉水湾"售楼中心一层平面布置图
PLAN 1:200

项目地址：厦门
竣工时间：2009年5月
设计师：陈方晓

代表作品（四）

星空

矗立在建筑工地边的售楼中心需要一个安静的氛围，就此设计师给售楼中心室内穿上神秘的深色的"衣服"。让来访者一进入便被精心营造的空间所催眠，坠入设计师所要的氛围。此时，色彩是空间的主角。深色营造的神秘感，让空间立刻安静。银色造型在深色空间里扮演着时尚的角色。红色造型点燃了人们心中暗藏的热情，穿梭的人流在红色映衬下如同沸腾的"太阳风"。

大厅天棚漫天的LED光纤灯头幻化成深邃的银河。镜面不锈钢制作的灯架和地面装饰线是流星的轨道，银色镜片装饰的圆形VIP洽谈区就是银河中的太空舱。VIP区悬挑出2m长的洽谈桌在灯光的衬托下营造出漂浮感，方形的发光体是降落在瓦砾上的外星来客。黑色的大理石地面反射着天棚的星空，我们置身在浩瀚的宇宙中。

Stars World

The houses selling center is near the construction site, so it need a quiet environment. Designers decorate the interior room by dark and mysterious colors in order to attract every visitor and make them accept designers idea, and hero of that room is color. Dark color create a kind of myth and a room can become very quiet. Models in silver color is a fashion performer and models in red color light the passion, so the hurried persons looks like a kind of solar wind.

A lot of LED lamps on the sunshade seems to form a galaxy. The stainless frame and decorative lines look like meteoric orbit and the round VIP chatting room in silver color is the same as space capsule. The table of two meters long is flowing under the light and the square shining objects is our guests from outer space, then the black marble reflects the sunshade, finally, we may feel that we are in the universe.

方国溪 Fang Guoxi

- 厦门辉煌装修公司 设计总监
- Xiamen Huihuang Decoration Company_creative Director

个人经历

- 福建省侨兴轻工学校
- 清华大学美术学院（原中央工艺美术学院）环境艺术系
- 意大利米兰理工大学室内设计管理硕士

个人荣誉

- 国家认证高级室内建筑师
- 中国建筑学会室内设计分会会员
- 中国优秀青年设计师
- 辉煌装修公司设计总监
- 亚太建筑师与室内设计师联盟理事
- 中国建筑装饰协会设计委员会委员

获奖荣誉

- 2004年获首届"东鹏杯"全国室内设计大奖赛最具创意奖
- 2006年获第二届"东鹏杯"全国室内设计大赛铜奖
- 2006年获"2006全国优秀医院设计师"
- 2007年度作品入围优秀建筑及室内设计作品展览"金外滩奖"
- 2007年荣获第三届中国国际设计艺术观摩展"最具潜力设计师"
- 2008年荣获"尚高杯中国室内设计大奖赛"优秀奖
- 2008年荣获"福建省首届艺术设计大展赛"铜奖
- 2008年荣获"亚太室内设计双年大奖赛"银奖
- 2009年荣获"亚太室内设计大奖赛"优秀奖
- 2009年荣获"第五届中国国际设计艺术博览会"资深设计师
- 2010年荣获中国建筑装饰协会全国有成就的资深室内建筑师

Experiences

- Qiao Xing School of Light Industry of Fujian Province
- Environmental Art Department of Tsinghua University
- Polytechnic University of Milan, Italy Master of Interior Design

Awards

- Senior Interior Architect
- China Institute of Interior Design Branch Member
- China's outstanding young designers
- Brilliant design director of the decoration
- Union Pacific director of architects and interior designers
- China Building Decoration Association Design Committee

Honor

- In 2004 the first "Dongpeng Cup "National Award for Most Innovative Interior Design Competition
- In 2006 the second "Dongpeng Cup "National Interior Design Competition Bronze Award
- In 2006 "2006 National Outstanding Hospital designer "
- In 2007 works of outstanding architectural and interior design finalist exhibition "Golden Bund Award "
- In 2007 by the Third China International Exhibition of Design Art to observe "the most promising designer "
- In 2008 was "still high interior design grand prix Cup of China" Award of Excellence
- In 2008 was "the first Art and Design Exhibition in Fujian Province Competition " Bronze
- In 2008 by the "Asia Pacific Interior Design Biennial Grand Prix"Silver
- In 2009 "Asia Pacific Interior Design Competition" Award of Excellence
- In 2009 "Art of the Fifth China International Fair " Senior Designer
- 2010 China Building Decoration Association, the country has achieved a senior interior architect

厦门是温馨的海滨城市，厦门城市不大，生活便利。生活、工作、压力也不那么大，所以厦门的设计有一种自然的小资"泡茶唇前、听涛看海"。中国设计近年发展很快，国内有很多设计师崭露头角，如何看中国设计？那得从国外看回来，那样看的话，中国设计又太"设计"了。我想是跟我们的思想观念有关，中国设计要在世界设计之林有一席之地的话，那得有值得人家喜欢的生活方式，说到底是文化，有值得传续的文化。

贵司主要从事哪方面的设计？最新的作品是什么？

· 主要从事医疗及地产的设计；最新的作品是北大国际医院、成都宋庆龄华西妇儿童医院。

您的设计灵感来源于什么？

· 设计的灵感要根据项目的属性、市场定位来进行分析，更多的是理性分析结合积累的视觉感受来完成的。

您从事设计工作多久了？您的设计理念是什么？

· 从事设计工作已十三年；设计主要关注的还是"怎样以人为本"以及"可持续发展"。

在设计过程中，您最注重哪方面的工作？对于多种设计元素有没有什么偏好？

· 注重了解项目背景资料，尽量做到业主需求+市场需要+"设计"的整合来达到设计的效果。对于设计元素没有偏好，我们都是根据项目定位来进行工作的，而且做公建项目较大较多，对设计会相对理性。

从前期接触项目、出概念、做方案到完成整个项目，觉得最大的困难是什么？

· 个人觉得最大的困难是如何保持效果的情况下，可以更节约资源的方式来营造。做为设计师在使用资源方面会较多，所以在使用材料方面我们慎之又慎，同样的效果，可以用不同的材料来表达。材料的碳排放、回收、制作成本、辐射等都是要结合考虑的，环境不好，装修得再好，品质也好不到哪里。

欣赏的或者对您影响最大的人是谁？抑或某种风格、思潮、理念？

· 我个人比较欣赏日本建筑师隈研吾在北京的作品"竹屋"，他所用的就是竹子，我想他考虑的除文化以外应该还有资源的问题。

怎么看待厦门设计？以及中国设计？

· 厦门是温馨的海滨城市，厦门城市不大，生活便利。生活、工作、压力也不那么大，所以厦门的设计有一种自然的小资"泡茶唇前、听涛看海"。中国设计近年发展很快，国内有很多设计师崭露头角，如何看中国设计？那得从国外看回来，那样看的话，中国设计又太"设计"了。我想是跟我们的思想观念有关，中国设计要在世界设计之林有一席之地的话，那得有值得人家喜欢的生活方式，说到底是文化，有值得传续的文化。

未来的公司会如何发展？您本人呢？

· 顺势而为，因势利导。

厦门御榕庄男子养生会所

设计意味着一段舒适的回忆。在刚接触"御榕庄"养生会所项目时,养生之道一触即发,引起了我们的思考。所谓的养生,对于人的体验是什么?当今工作压力很高,生活琐事很多,快乐却很少的状态下,人人是否都需要找回属于自己的安静?譬如回归自然。

思绪追溯着大自然的天籁灵气。犹如美丽的殿堂,树香,鸟鸣,水声,光影……道法禅思,听觉,嗅觉,味觉,触觉重新唤起的喜悦。生命在此得到沐浴,宛如新生,享拥人间最美好的时光。

Men's Health Club Royal Banyan Tree

Design means a comfortable memories. In the new to the "Royal Banyan Tree" health club projects, and good health to explode, causing our thinking. The so-called health, for the human experience is what? Work pressure is high today, a lot of trivia, happy little state, if everyone needs to find their own quiet? For example, return to nature.

Thoughts back the sounds of nature spirit. Like a beautiful palace, trees, birds, water, sound, light and shadow ... Road Act meditation, hearing, smell, taste, touch, renewed joy. Be bathed in this life, open like a newborn, the best time to enjoy human owner.

项目地址:厦门
设计师:方国溪
设计单位:厦门辉煌装修公司

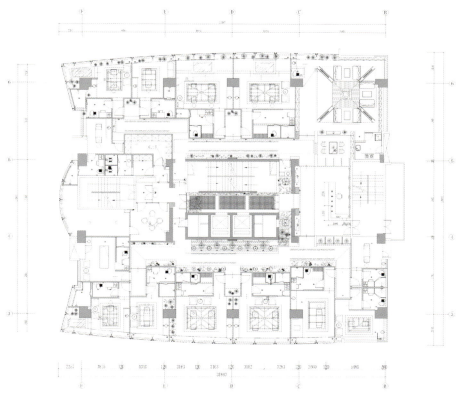

/ 平面布置图 /

厦门五缘湾样板房

本样板房位于五缘湾，利用得天独厚的地理位置，结合该区域的功能规划，本设计的主题正是营造出一种舒适、休闲、高品质的游艇家居生活，使人幻想迷恋大海所拥有的神秘力量、游艇所蕴涵的海洋文化，让人回到家就有一种享受大自然、阳光、海滩,分享快乐的喜悦心情。主要以白色调为主，利用各种材质的变化将游艇的元素设计运用的淋漓尽致，使人们生活在当中，有身临其境的感觉，使室内外完全融合为一。

Wuyuan Bay Open Houses

The model room is located in Wuyuan Bay, the use of geographical location, combining the functions of the regional planning, the theme of this design is to create a comfortable, casual, high-quality yachts home life, people have the sea fantasy obsession Mysterious power, yacht marine culture implied in, people have a home to enjoy nature, sun, beach, happy to share the joy. Mainly white tone-based, using a variety of material changes to the elements of yacht design using the most of, so that people living in these, immersive feeling, so is a fully integrated indoor and outdoor.

项目地址：厦门
设计师：方国溪
设计单位：厦门辉煌装修公司

代表作品（二）

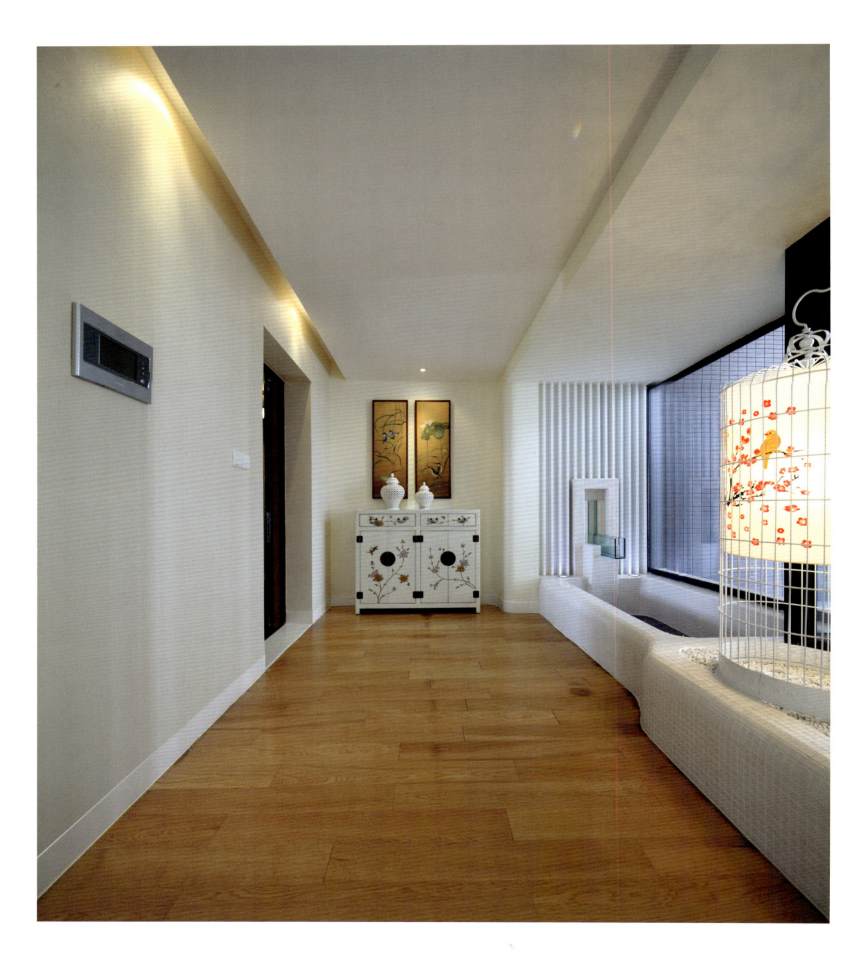

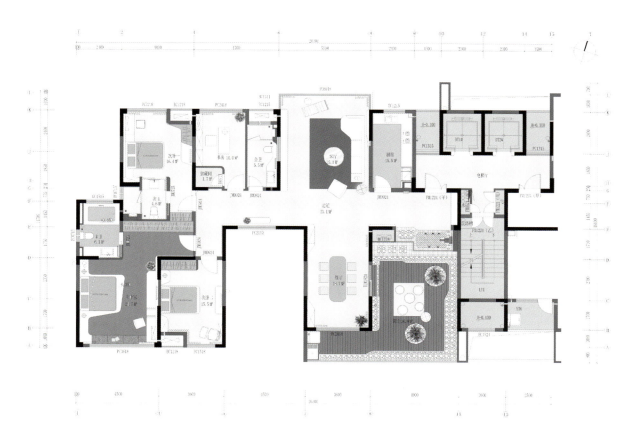

/ 平面图 /

方令加 \Fang Lingjia

- 三佰舍室内设计顾问有限公司_设计总监
- Sanbaishe Interior Design_Creative Director

个人经历

- 1981年-1993年　农民
  1993年-1997年　工人
  1997年-2000年　学徒
  2000年-2003年　无业
  2003年-2007年　喜马拉雅设计师
  2007年组建三佰舍室内设计顾问有限公司

Experience

- 1981-1993　Farmers
  1993-1997　Workers
  1997-2000　Learner
  2000-2003　Unemployed
  2003-2007　Himalayan Designer
  2007 Sanbaishe Interior Design Consultants Ltd.

也许80年代的人都有这样的梦想，有一所房子，不一定面朝大海，但是至少也在山野之间，回归乡村和自然。我是在农村长大的，在城市里觉得拥挤、压抑，比较渴望回到农村。

贵司主要从事哪方面的设计？最新的作品是什么？

· 主要是小型的商业空间，比如休闲场所、餐厅会所、样板房等。最近有好几个小空间即将完成。

· 今年我第一次接触样板房项目，从设计到软装，它非常能锻炼设计师。在项目没有开始之前，开发商已经有很多经验，包括很多细节、各种规范和技巧，这能让设计师学到很多。

您的设计灵感来源于什么？

· 我的灵感来源于误解。误解一个事物，然后发现不同于常规的思路，用全新的角度诠释。误解也可以赋予事物的一种转变、一次新生。比如，如果眼睛不好的人来做设计，我想他一定做的很不错。

您从事设计工作多久了？可以简单介绍一下吗？

· 从事设计的时间并不长，大概五年，现在我仍旧再学习、不断成长。怎么让客户接受是我一直以来的不断面对的难题。即使你用极大的心血做出设计方案，客户不认可不接受，也是功亏一篑。也许这和经验、名气相关。所以现在我尽量会参加比赛，更加注重宣传。

您在做设计的过程中，觉得最大的困难是什么？

· 我觉得时间不够用。我还处于学习阶段，发展并不算快。我希望可以稳步前进，每个项目我都想做得更精细、更完美，但是业主的时间是有限的，所以常常感到时间不够用。所以现在我还没有接规模比较大的项目，一个是经验，再就是自己还需要学习。比如酒店项目，需要建立一个很大的团队，也需要重新学习很多酒店设计的知识，我需要更多的时间，但是也许业主并不一定能给予我这样的机会。

您欣赏的或者对您影响最大的人是谁？抑或某种风格、思潮、理念？

· 基本上大师级的建筑师和设计师都蛮欣赏的，安藤忠雄、扎哈。国内设计师孙天文的个性我也非常欣赏。

平时有什么爱好吗？

· 我喜欢到各地去收集古玩、旧货。厦门有7个面积很大的古玩城，我喜欢去捡漏、淘宝，在里面搜到很多好东西。我现在喜欢看很多建筑设计方面的书籍。

未来的公司会如何发展？您本人呢？

· 以前看徐悲鸿的传记，说他在国外学习的时候每天都去美术馆临摹名画，因为没有什么钱，常常吃不饱，后来他发现吃不饱的时候画得最好。对于公司的经营，也许不一定满足温饱，但是尽量不饿肚子就好（笑）。吃半饱的时候可以保持清醒，也许会将设计做得更好。我在向景观建筑的方向调整。如果很有能力应该是建筑景观室内一起设计。

· 也许80年代的人都有这样的梦想，有一所房子，不一定面朝大海，但是至少也在山野之间，回归乡村和自然。我是在农村长大的，在城市里觉得拥挤、压抑，比较渴望回到农村。

一尊皇牛

餐厅是以经营肥牛为主的火锅料理。空间分三层,整体运用不多的材料种类以追求用餐环境应有的舒适度,并结合不同的空间处理,保证三层分别以各自的感受展示给客人,配合合理的灯光及艺术和自然的大体积装饰品,以使人产生冥想和回忆

Yizun Huangniu Restaurant

Restaurant is a pot with main dishes beef. Space divided into three layers, the overall use of the few types of materials due to the pursuit of comfort dining environment, combined with different spatial processing, to ensure the feeling of three-tier display of each of their guests, with reasonable lighting and art and nature. The large volume of decorations, to engender meditation and memory.

项目地址:厦门
设计师:方令加
设计单位:三佰舍室内设计顾问有限公司

代表作品(二)

清汤餐厅

餐厅的环境，和料理一样，清淡、朴素，是现代的，也是中国的。

大空间的关系根据功能需求分隔为不同大小的各个方形空间，型体上并无做多余的修饰，以使整体框架是纯粹和现代的，墙角的黑边以使墙面不会被大面的白色溶化，保护了墙角，也使空间的骨架上带有最为简洁的中国味道。把对比强烈的或老或新的家具及摆件置入空间，使传统和现代在干净的空间中交错游走。

Qingtang Restaurant

Restaurant environment, like food, light and simple, is also a modern and also is China.

The relationship between the large space separated according to functional requirements for different sizes of each square space, type of body modification is not made redundant, so that the overall framework is a pure and modern, black side wall so the wall will not be large areas of White melt and protect the wall to make room for the skeleton is also the most compact with Chinese flavor.The contrast or old or new furniture and ornaments into space, traditional and modern in a clean space staggered walk.

设计师：方令加
竣工时间：2010年9月
项目地址：厦门
项目面积：480m²

代表作品（二）

/ 一层平面图 /

/ 二层平面图 /

胡若愚 \ Hu Ruoyu

- 厦门喜玛拉雅设计装修有限公司_设计总监
- Co., Ltd. Xiamen Himalayas_Creative Director

个人经历
1992年厦门大学建筑系本科毕业留校
1998年创立厦门喜玛拉雅设计装修有限公司

Experience
In 1992, Xiamen University, Department of Architecture graduate school to teach
In 1998, the creation of design and decoration Co., Ltd. Xiamen Himalayas

厦门之前业界的联系比较少，现在慢慢多了。相互之间的交流，可以促进相互学习。我很喜欢和年轻的设计师在一起，从他们身上我也学到了很多。

您从事设计工作多久了？公司成立多久了？规模如何？

· 我毕业后就开始做设计，到现在有十七八年了。厦门喜马拉雅设计装饰公司是1998年成立的，设计师有60~70人左右。按区域分为三个设计部门，分别是北京、成都和厦门三个中心区域，设计、施工业务都有。

您的公司更加全面多样，那么在设计中，照明和软装设计是怎么完成的？会同专业的公司合作吗？

· 照明设计是我们自己完成的。但是软装部分的建设并不成熟。公司有侧重这方面的设计师，但是没有专业的软装设计师。之前也同厦门当地的软装公司配合，但是我们并不满意。软装设计师其实要求非常高，它要求很高的设计品位。可以很快培养出一个室内设计师，但是很快培养出一个品味好的软装设计师是非常困难的。

这十几年来，您觉得工作中遇到的最大的困难是什么？

· 工作本身的困难并不多，困难都在于人。

· 从个人角度看，一个人的定位非常关键。我们在做设计的过程中或多或少都会涉及到施工，施工非常琐碎、复杂，和设计是不一样的工作。不做施工，设计的实现度又会打折扣；承担了施工任务，又会占用很多时间，无法静心设计。角色的转换，就是人必须面对的困惑，更是一个自我定位的问题。两者一定要权衡好，对自己有一个准确的定位。这几年我在施工方面花了比较多的精力，不过我也还在做设计。

· 从整体角度而言，就是公司的管理问题。公司发展中受到各种条件约束，发展相对平稳，但是年轻人的学习能力和个人发展也许是很快的。一旦公司的发展跟不上个体的发展，换言之就是限制了个人的发展，就会造成人才的流失。厦门是一个处于发展阶段的中小城市，有很多的小空间项目，利于设计师独立。因此，必须要更好的解决公司管理问题，比如采用合伙人的制度等。

对设计教育有什么看法？如何对公司的设计师进行再教育？

· 学校教育灌输的知识导向使得学生都希望成为大师。但在现实中，社会并不需要那么多大师。如果自己定位过高，和现实有很大的差距，个人将会痛苦，企业也很痛苦。再次，首先你必须成为一个合格的设计师，才能进入企业，从事设计工作。我曾经是厦门大学建筑系的老师，讲课都是教一些很基本的理论知识，实践和应用还是靠学生个人。设计专业更加脱离现实，专业混乱，师资力量不够。教设计的老师自己都没有做过设计，没有来自第一线的知识和经验。设计行业的变化是很快的，如果不在第一线，就会脱离现实。所以企业还需要很长的培训时间，增大了企业成本。我们会安排已经有一定经验的设计师接受培训和再教育。

怎么看待厦门设计？

· 厦门之前业界的联系比较少，现在慢慢多了。相互之间的交流，可以促进相互学习。我很喜欢和年轻的设计师在一起，从他们身上我也学到了很多。

厦门五缘湾样板间

在不大的居家格局内，尝试在空间的设计整合中体现东方的审美情趣。空间系列围绕半室外的内凹阳台而展开，以此为中心完成空间的层次节奏变化和内外交融：海景→书房→过道→内阳台→小区内院是一条轴线，相垂直的是厨房→餐厅电视墙→餐厅→内阳台→主卫淋浴区的另一条轴线。内阳台既是第二会客厅，也丰富了餐厅的景观视野，同时又是饱读之余养目静心的所在，而泡澡之后推门而出在躺椅上放松身心也是一种享受。

Wuyuan Gulf Show Flat

Pattern in the small home, the attempt to integrate the design of the space reflect the aesthetic taste of the East. Rhythm changes and the level of space inside and outside the blend: sea - the library - hallway - the balcony - the district hospital is a axis perpendicular to the kitchen - TV Wall - Restaurant - Terrace - another axis of the shower area. Only the second sitting room balcony, but also enriched the landscape view restaurant, is also well-read, apart from raising project where meditation.

项目地址：厦门
设计师：胡若愚
设计单位：厦门喜玛拉雅设计装修有限公司

代表作品（一）

/ 平面图 /

为强化空间的趣味性和层次感，书房和主卫均采用落地木格栅和纸折叠门，而客厅沙发背景则是黑钢框凹凸冲砂木块移门，推开后，客厅和书房的空间融为一体。

To strengthen the sense of space and levels of interest, study and main floor are made of wood grille guard and paper folding, while the living room sofa background is sliding doors, open living room and study room for post-integration.

主要用材：客餐厅主卫墙地面为意大利木纹石，走道墙面及卧室木地板为柏丽复合强化木地板，内阳台为绿可木和火烧水洗石。整体风格体现现代东方情调，时尚简约中隐约着东方的飘逸大气。

To strengthen the sense of space and levels of interest, study and main floor are made of wood grille guard and paper folding, while the living room sofa background is sliding doors, open living room and study room for post-integration.

成都金沙鹭岛售楼处

密林中一片石墙斜穿，毛面、哑面、光面质感的黑色花岗岩混搭；一面水镜，几点星灯，几段木化石横卧，几阶木平台错落；为达至室内外相融，建筑的围护为无框钢化玻璃，支撑钢构屋顶的钢柱也尽量纤细，钢构屋顶仿佛漂浮在空中。接待台为花岗岩方料切凿，地灯映衬下自然断裂面的力度感更为强化；围绕着接待台的是几片横向延伸的木作承板，任意角度的黑钢斜片穿插其间。吧台上天花延伸而下的木格架罩着高低错落的云石光块；视听墙为简化的现代壁炉，呼应的是入口黑色石墙的质感组合；在大厅与后场之间，由斜墙延伸隔出的过道空间内，洗手台成为视觉焦点；一片镜面将椭圆形花岗岩方料一剖为二，门字形木格架又隔出几分私密；地面皇室木纹大理石从室内延伸至户外平台与天棚冲砂木的凹凸肌理相呼应。

Egret Island Sands sales offices

Forests in a wall, rough surface of black granite; mirage, light, wood technology, wood platform. Building Envelope as frameless glass, steel roof supported by slender steel columns are also possible, steel roof, as if floating in the air. Front Desk for the granite square stock cut chisel, to light against the background of a sense of the natural strength of the fracture surface is more enhanced; around the reception desk is a few pieces of horizontal wood for the board, any angle oblique black steel sheet during interludes. Ceiling extension of the bar down the level of the wooden frame covered with scattered marble; audio-visual wall to simplify the modern fireplace, stone walls echoed the entrance of black; in the hall and back between the extension, separated by a sloping wall of the aisle space, sinks and a visual focus: Mirror of granite materials for a split into two parties; the ground extending from indoor to outdoor marble.

项目地址：成都
设计师：胡若愚
设计单位：厦门喜玛拉雅设计装修有限公司

最新作品（二）

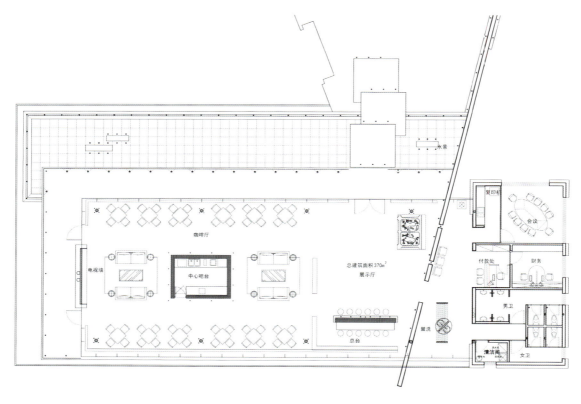

/ 平面图 /

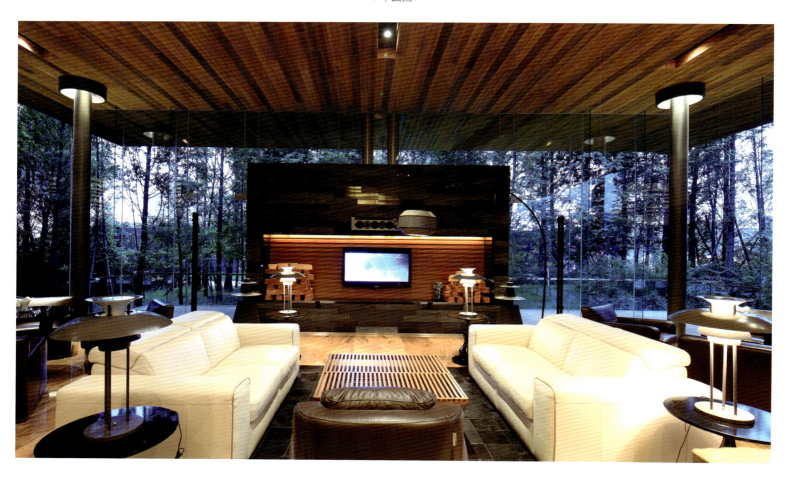

黄少雄 \ Huang Shaoxiong

- 同3组_设计总监
- Tong3_Creative Director

个人经历
- 同3组设计总监
  厦门东方设计研究院 副院长
  同济大学工学学士
  清华大学建筑学院结业
  意大利米兰理工大学室内设计硕士生

获奖荣誉
- 亚太酒店设计协会（APHDA）理事
  中国建筑装饰协会会员
  作品"金门湾大酒店"获2008年全国建筑工程装饰奖

Experience
- Tong3 Creative Director
  Design Institute of the East of Xiamen Deputy dean
  Bachelor of Engineering, Tongji University
  Closure of Architecture of Tsinghua University
  Polytechnic University of Milan, Italy Interior Design Master

Awards
- Asia Pacific Hotel Design Association (APHDA) Director
  China Building Decoration Association
  "Golden Gate Bay Hotel" won the 2008 National Construction Decoration Award

厦门设计也是中国设计的一个缩影，设计任务很多，大家都很忙，有很大的发挥锻炼空间，我们都想向国外的领先设计企业看齐，在概念设计、细部设计、标准制图、物料选配方面都要努力和提高，差距会越来越小的。

贵司主要从事哪方面的设计？最新的作品是什么？

· 我们主要从事酒店、办公和地产配套项目的室内设计，景观和建筑设计也做。新近刚建成的项目有：一家上市公司的小型会所酒店、一家首批创业板的公司总部、一家精品形象店、一家大型工厂的新办公楼、几个地产项目的售楼中心。项目是建成了，但我们觉得还称不上作品。

您的设计灵感来源于什么？

· 设计灵感主要来源于生活体验。我觉得体验的积累很重要，体验的内容包含很多方面。能激发灵感的体验有可能是同类空间的感受，可能是一幅画、一处自然景观，也可能是一篇文章、一段历史。

您从事设计工作多久了？您的设计理念是什么？

· 1990年大学毕业后就开始做室内设计，同时也做施工管理。我认为室内设计是一项建构使用空间和创造室内气氛的工作，是兼具理性与感性的工作。我会更重视建筑空间设计在室内气氛营造所起的作用。

在设计过程中，您最注重哪方面的工作？对于多种设计元素有没有什么偏好？

· 一个设计是由几个阶段构成的，每个阶段都很重要。同项目业主的沟通、概念设计、施工图设计、选材、施工设计管理等，少一个环节控制都不可能有好的设计成果。我没有特别偏好哪种设计元素，应根据具体的项目要求来选择空间表达语言，尽可能现代简洁些。

前期接触项目、出概念、做方案到完成整个项目，觉得最大的困难是什么？

· 不同的项目面对的问题不同，困难也不一样。我觉得最大的困难是时间问题，大部分项目的设计周期都很短，没有充足的时间去优化方案，没有充足的时间去选择比较材料，没有充足的时间去推敲节点构造，很遗憾。

欣赏的或者对您影响最大的人是谁？抑或某种风格、思潮、理念？

· 国际大师很多，我比较欣赏华人设计师如贝聿铭、李裕棠。他们将建筑和文化结合得很好，又很有时代感。

怎么看待厦门设计？以及中国设计？

· 厦门设计也是中国设计的一个缩影，设计任务很多，大家都很忙，有很大的发挥锻炼空间，我们都想向国外的领先设计企业看齐，在概念设计、细部设计、标准制图、物料选配方面都要努力和提高，差距会越来越小的。

未来的公司会如何发展？您本人呢？

· 我们目前不想再扩大规模，希望把服务的客户项目做好，把设计质量做好，把团队建设好。我喜欢设计，设计已是生活的一部分。要做好设计，还要继续学习。想得不太多，生活、设计、学习就这么一直过下去了。

国贸蓝海F型样板房

本案是一高档海景楼盘样板房，设计概念源自地中海沿岸的一幢住宅建筑室内，不是具象地去模仿，而是用完全现代的设计语言来诠释。

希望通过设计,让居所主人放松自在地生活在由天然材质和海景编织的环境里。

Blue Ocean F-Trade Show Flat

This is a show flat for a high-end sea-view real estate. Its concept of design originates from the interior decoration of a house building along the Mediterranean Sea. It is not the copy of visual details but is the interpretation of the idea by modern design language.

The aim of this design is to let the house owner live in a relaxing environment composed of natural materials and sea-view.

项目面积：225 m²
项目地址：厦门市五缘湾
设计师：黄少雄
摄影：吴永长

代表作品（一）

/ 平面图 /

同3组办公楼

同3组办公楼是一幢二层的独立建筑，建筑里外进行了一致的设计改造。目的是给设计师创造一个充满自然光线和自然景观的舒适工作空间。

自然光是通过"调理"后进入室内的。一层临街的落地窗外种满了竹子，光穿过竹林进入室内。与相邻建筑一侧的窗采用玻璃砖砌筑，光有了色彩和流动的纹理。工作区域的落地窗选用LOW-E玻璃配竹百页帘，光不再炽热还可手动调节。午后的一个时段，光还会将室外的树影打进室内，这时空间显得异常的生动。

Office Building With Tong3

Tong 3 office building is an individual two-storey building, harmonically designed and renovated interiorly and exteriorly, aimed to create for the architect a comfortable working space which is full of natural light and scenery.

The natural light enters the room through some "adjustment". On the front side, the bamboos are planted along the roadside just outside the French Windows of the first floor. The light enters the room through the bamboo grove. The backside faces the nearby buildings. Here the glass tiles are used for windows, giving color and floating veins to the light. In the working area, LOW-E glass is used for French windows to match the bamboo blinds. The rays of sun is not burning but can be adjusted by hands. Some time after noon, the light will reflect the outdoor shadows of the trees into the room, making the space look especially vivid.

项目面积：300 m²
地址：厦门市吕岭路272号
设计师：黄少雄
摄影：吴永长

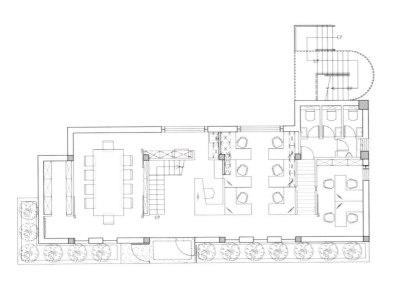

/ 一层平面图 /

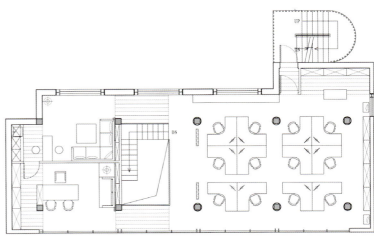

/ 二层平面图 /

黄振耀 \ Huang Zhenyao

- 东峻设计顾问有限公司_设计总监
- Dong Jun Design Consultants Ltd_Creative Director

个人经历

- 东峻设计顾问有限公司设计总监
  意大利米兰理工大学室内设计管理学硕士
  清华大学酒店设计高级研究生班毕业
  IAI亚太建筑师与室内设计师协会福建分会副理事
  IDA国际设计师协会分会副会长
  中国建筑协会室内设计分会资深会员

获奖情况

- 获得IDA 2008国际建筑景观室内设计大奖赛样板房类特等奖
  获得IDA 2008国际建筑景观室内设计大奖赛样餐饮类金奖
  获得2009国际建筑与室内设计节金外滩奖提名奖
  获得2009国际建筑与室内设计节金外滩奖4个入围设计奖
  获得17届香港亚太室内设计大赛酒店空间前十名
  获得2009中国室内空间环境艺术设计大赛酒店类三等奖
  获得2010年中国（上海）国际建筑及室内设计节金外滩"最佳概念设计奖"
  获得2010"INTERIOR DESIGN China 酒店设计奖——最佳酒店设计奖"

获奖荣誉

- 作品录入《2009优秀室内设计师》
  《中国创意界》
  《17届香港亚太室内大奖入围及获奖作品集》等相关作品集

Creative Director

- Dong Jun Design Consultants Ltd Design Director
  University of Milan, Italy Master of Interior Design Management
  Hotel Design, Tsinghua University High graduate class graduates
  IAI Pacific Architects and Interior Design Association, vice director of Fujian Branch
  Branch of IDA, vice president of the International Society of Designers
  Interior Design Branch of China Construction Association Fellow

Awards

- IDA2008 international architectural landscape model room Interior Design Competition Grand Prize category
  IDA2008 International Architecture Landscape Interior Design Competition Award categories like food and beverage
  2009 International Festival of Architecture and Interior Design Award Gold Award nomination Bund
  2009 International Festival of Architecture and Interior Design Awards 4 Gold Bund Design Award Finalist
  The 17th Asia Pacific Interior Design Competition Hong Kong hotel room for the top ten
  2009 China Interior Design Competition Art Space Environment Third Class Hotel
  2010 China (Shanghai) International Building and Interior Design Festival Bund "Best Concept Design"
  2010 "INTERIOR DESIGN China Hotel Design Award——Best Hotel Design "

Honor

- Works of entry "2009 Outstanding Interior Design"
  "creative sector in China, "
  "17th Hong Kong and the Asia Pacific Interior Award winning set of short-listed"and so on

厦门是很幸福的城市，厦门有很多很出色的设计师，也有好多朋友都很有个性，在做原创方向的研究，希望业主也能有这方面的思维和支持，那大家就有可能看到厦门的魅力了。中国需要"设计"，需要建立在我们自己人文基础上的设计，设计方面我们的业主和设计师都不必模仿任何国家，研究一下我们五千年的文化历程，用当代的思维表现出来，就足以成为世界设计的方向标了。

贵司主要从事哪方面的设计？最新的作品是什么？

· 我公司主要从事个性建筑，地产项目和个性化酒店、餐厅、办公等项目的策划和室内设计。最新的设计很多，但大部分还在施工阶段，有兴趣可关注公司网站：WWW.ET-CN.COM。

您的设计灵感来源于什么？

· 设计灵感大多来自对生活的感悟，生命的思考，还有自己看到的、听到的能给自己留下深刻印象的事物。

您从事设计工作多久了？您的设计理念是什么？

· 从事设计工作有十几年了，设计理念和公司的座右铭一致："观念改变思维，概念决定高度"。

在设计过程中，您最注重哪方面的工作？对于多种设计元素有没有什么偏好？

· 在设计过程中最注重的是前期准备工作，先听客户对整个项目的诉求，对项目做一个方向性评估和诊断，站在受众的角度去考虑使用者的感受细节，站在经营者的角度分析成本与效益，才开始结合自己的专业水平进行深化。设计元素方面并没什么偏好，对自己创造的元素比较有兴趣。

从前期接触项目、出概念、做方案到完成整个项目，觉得最大的困难是什么？

· 最困难的事是超越自己，因为每个项目的性质都不同，成功演员最珍贵的是塑造不同角色性格和内心世界，演什么像什么，设计师最难的是如何超越自己的思维模式。

欣赏的或者对您影响最大的人是谁？抑或某种风格、思潮、理念？

· 欣赏的人很多。

怎么看待厦门设计？以及中国设计？

· 厦门是很幸福的城市，厦门有很多很出色的设计师，也有好多朋友都很有个性，在做原创方向的研究，希望业主也能有这方面的思维和支持，那大家就有可能看到厦门的魅力了。中国需要"设计"，需要建立在我们自己人文基础上的设计，设计方面我们的业主和设计师都不必模仿任何国家，研究一下我们五千年的文化历程，用当代的思维表现出来，就足以成为世界设计的方向标了。

未来的公司会如何发展？您本人呢？

· 我们公司要的是质化，不要量化，没想把公司做的规模有多大，今年加入加拿大E&T国际设计顾问，人员结构上更精良点，设计费看能否拉高一点。让大家越来越放松，但钱要越赚越多，作品越来越对得起大家。这就是我的发展方向了。

PARK.酒店

PARK.HOTEL位于素有万国建筑博览中心的度假小岛鼓浪屿。风光秀丽，人文风情极其丰富。可惜该项目是一栋失忆的小楼，两年前被开发公司重建，把美好的东西给埋葬掉了，没有故事多少是一种遗憾。但曾几何时是谁为鼓浪屿谱写这么多动人的故事，我在考虑今天的我们是否有能力为这么美丽动人的小岛写一段属于当代情结的篇章。300m²实用面积的小楼，多少有点难度，还好有漂亮的花园。经过实地的勘查300m²扣除接待空间和作为酒店所需的功能分区外最多可作8间客房。

PARK.Hotel

PARK.HOTEL located on the resort island Gulangyu. Beautiful, extremely rich cultural customs. Unfortunately, the project was developed two years ago, the company rebuilt the good things to bury, no story of how much is a regret. I was thinking today we have the ability to write such a beautiful island is a modern complex chapter. 300m² of usable area of the small building was somewhat difficult, but fortunately there are beautiful gardens. After deduction of the field survey 300m² reception space and the required function as a hotel outside the district can do up to 8 rooms.

项目面积：300 m²
地址：鼓浪屿复兴路6号
设计师：黄振耀
竣工时间：2009年6月

代表作品（一）

8间客房的故事要怎么写？能否有8种不同的设计理念，表达8种不同的情感生活体验。情感才是最重要的。作品的物理价值我们大家做的太多了，应该多试玩一些体现情感价值的东西。我为8个房间取了8个包含8种不同情感元素的名字。根据不同的情感故事，通过不同的色彩、灯光、材质、工艺的演变进行诠释，让使用者产生相应的情感共鸣。

8 rooms how to write the story? Whether there are 8 different design concept, the emotional expression of 8 different life experiences, emotion is the most important, the physical value of the work we do too much, should try to play some things of sentimental value reflected. I fetch the name of the element with different emotions, depending on the emotional story, through different color, lighting, material, interpretation of the evolution of technology, allowing users to generate the appropriate emotional response.

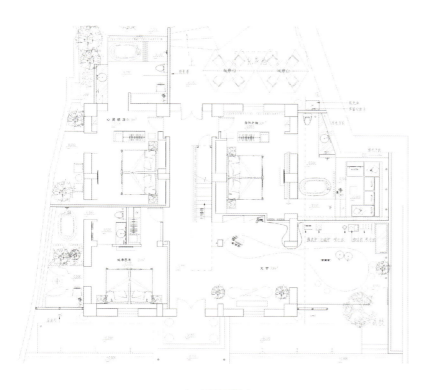
/ 一层平面图 /

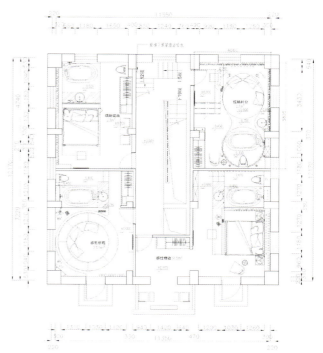
/ 二层平面图 /

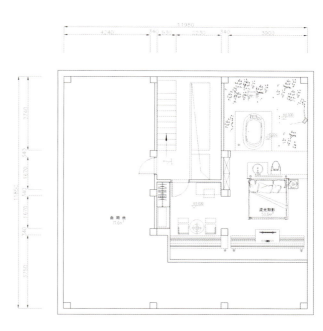

/ 三层平面图 /

中骏天峰会所

设计围绕着如何在长21m高7.6m的局限性空间中满足销售会所的刚性功能需求的前提下,通过造型灯光、色彩、材质等设计需要来彰显主题定位(摄政王钻石)的空间气质。

Zhongjun Tianfeng Club

Under the guideline of realizing the basic functions of the sales salon in a limited space which is 21m high and 7.6m long, this design demonstrates the noble quality of the theme (Regent Diamond) through special combination of the lights, colors and materials.

设计师:黄振耀
设计单位:东峻设计顾问有限公司

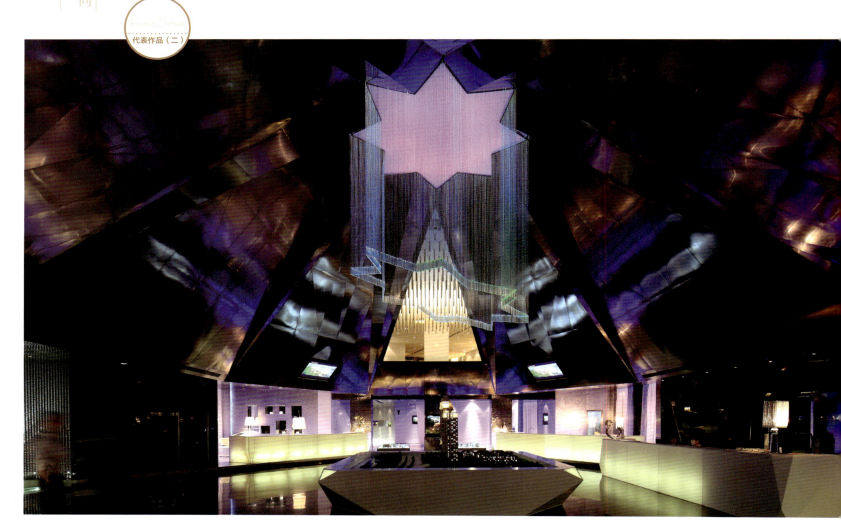

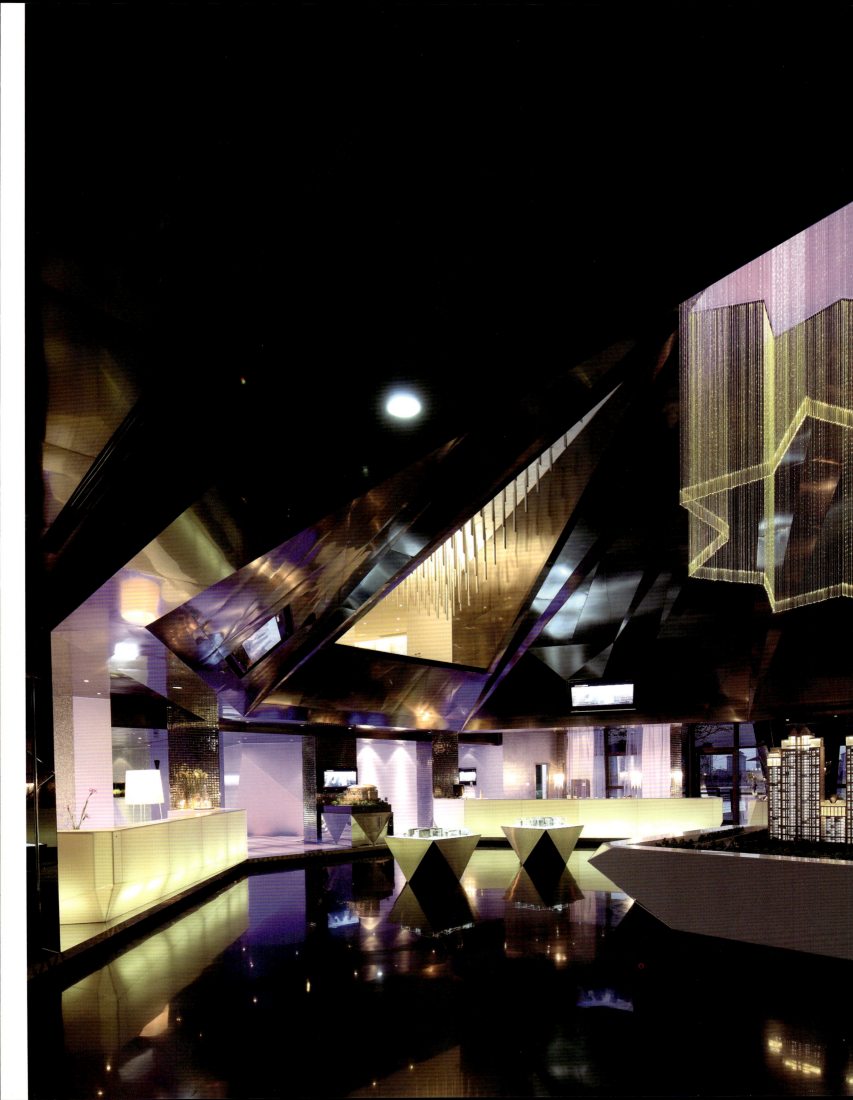

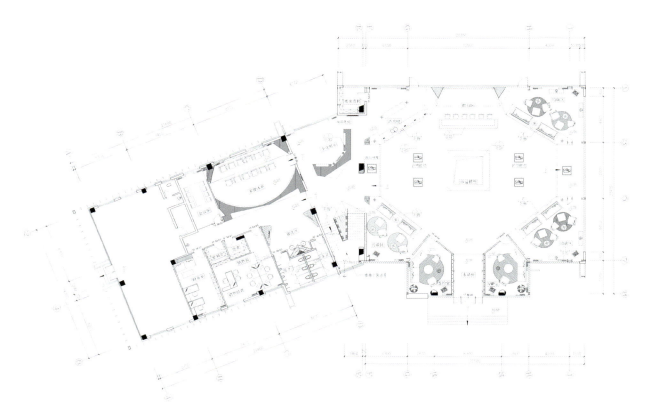

/ 平面图 /

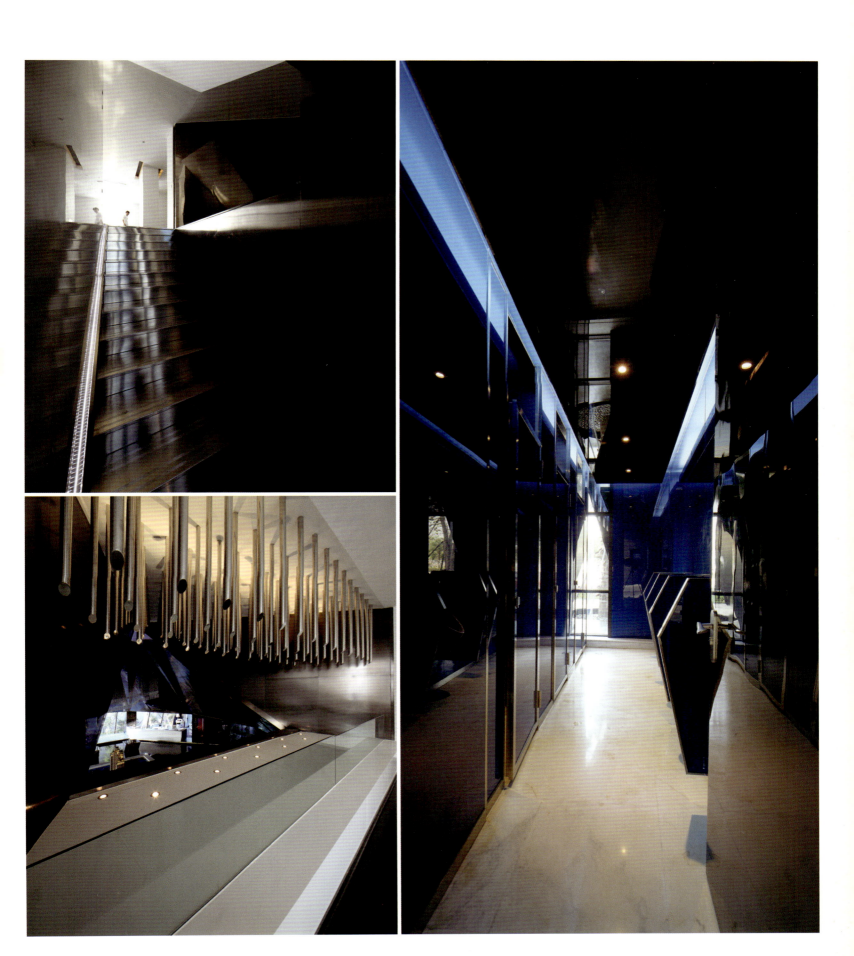

丽湾优家售楼处

本案现代感十足的造型与外立面的处理交相辉映，外部富于形式感的灯带布置强化了整栋建筑的动态线条，给人以刺破云霄的冲击力，建筑精神尽显。位于建筑中心的一石水小景更是别有洞天，室内设计方面延续了现代感与奢华气质，石材的运用于建筑外立面的木材相映成趣，值得细细品味。

Liwan Youjia Sales offices

Case, modern styling and full treatment of each other facades, external light with a rich sense of form layout reinforces the dynamic lines of the entire building, giving the impact punctured the sky, building the spirit of filling. Center for Architecture is located in a small stone water scenery is amazing, the interior design continues the modern and luxurious qualities, the stone facades of wood used in building a sense of exist side by side, it is worth a careful look.

设计师：黄振耀

设计单位：东峻设计顾问有限公司

代表作品（三）

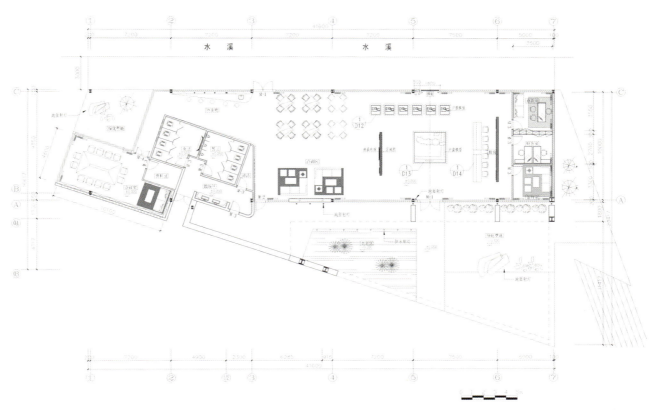

/ 平面图 /

赖云舟/Lai Yun Zhou

- 厦门大韵天成设计有限公司_设计总监
- Dayun Tiancheng Design Co., Ltd._Creative Director

个人经历
- 中国建筑学会室内设计会员
  IAI亚太建筑师与室内设计师联盟理事会员
  厦门大学艺术学院美术系毕业
  从事室内设计16年
  2007年成立厦门大韵天成设计有限公司，现任设计总监。

获奖情况
- 2009年获IAI亚太室内设计精英邀请赛方案类金奖
  2009年获"金外滩"最佳概念设计优秀奖
  2008年尚高杯中国室内设计大奖赛佳作奖
  2008年获第二届亚太室内设计双年大奖赛提名奖
  2007年获汉斯格雅福络创意设计赛卓越设计奖

Experience
- Member of China Institute of Interior Design
  Architects and interior designers IAI Union Pacific board members
  Xiamen University, Graduate Institute of Arts
  16 years in interior design
  In 2007, the establishment of Dayun Tiancheng Design Co., Ltd, design director.

Awards
- 2009 IAI Asia Pacific Interior Design program of the elite class of Gold Invitational
  2009 "The Bund"Best Concept Design Excellence Award
  2008 Cup of China is still high masterpiece Interior Design Competition Award
  2008 Second Asia-Pacific Interior Design Biennial Award nomination Grand Prix
  2007 Hansgrohe Design Excellence Award for Creative Design Competition

保持激情是不懈的动力,我觉得除了自己对于设计的热爱,还有就是设计本身就充满创意、要求不断的求新求变,我对不同的空间,自然就有一种新的激情,这种激情会一直激励我产生新的创意。当然在设计中还必须考虑到业主的需求,如果脱离了生活,脱离了业主的需求,再好的创意都是无意义的,设计师就是艺术性、专业性的完成业主的品味空间。

您公司取名"大韵天成"这是否代表着您对于设计的追求"大韵天成,完美追求"?

· "大韵天成"基本可以概括我对设计的追求。好的作品应该是浑然天成的,自然而然有我们的思想和创意在里面,我对于设计图纸也力求完整和完美,对客户我宁可花更多的时间做出一个好的案子,也不愿意去敷衍应付。每个设计案子都是全身心地投入,一定要自己觉得满意了,才把作品呈现出来。

您如此浑然天成的设计灵感是从何而来呢?有哪些文化影响到您的设计或者体现在您的设计中?

· "灵感来源于生活细节,文化底蕴是设计的灵魂。"中国文化应该是影响最大的,因为文化的熏陶和根生的环境,在设计中不自觉地表露出来。还有就是西方现代艺术如绘画、建筑带来的冲击,让我思考着设计的精神和方向。

您这几年来在设计道路上不断的取得丰硕成果,可否分享一下您的心得?

· 保持激情是不懈的动力,我觉得除了自己对于设计的热爱,还有就是设计本身就充满创意、要求不断的求新求变,我对不同的空间,自然就有一种新的激情,这种激情会一直激励我产生新的创意。当然在设计中还必须考虑到业主的需求,如果脱离了生活,脱离了业主的需求,再好的创意都是无意义的,设计师就是艺术性、专业性的完成业主的品味空间。

惠安聚龙小镇售楼处

翠绿环抱、白马仙居、云蒸雾绕……神仙眷侣的梦境空间。

本案是位于青山绿水间的商场建筑售楼处，只可惜设计空间无法延展到水淌风拂的户外，欲置写实造景于室内，反倒显得区区小景无法展现自然造化，只有升华成一种意境才能让顾客体悟到耳目一新的境界。因此，整体设计意在创造出一个飞离小景而氤氲出的意境，空间的每一个角落都渗透着浓厚稠密的梦境感。

Sales Offices Of Hui'an Julong Town

Surrounded by green, cloud steaming fog around the ...... dream space.

This case is located between the mountains of marketplace sales offices, but unfortunately the design space can not extend to the outdoor wind blowing the water drips, realistic scenery in the room, hand, can not seem to show little scene mere natural good fortune, only sublimated into a mood to let the customer realized that a fresh state. Therefore, the overall design is intended to create a little scene and dense fly out of the mood, every corner of space are permeated with a strong sense of dense dreams.

项目面积：2000 m²

设计师：赖云舟

主要材料：水曲柳、石材、绿草皮、不锈钢管、方管、白石子、白色枯树

代表作品（一）

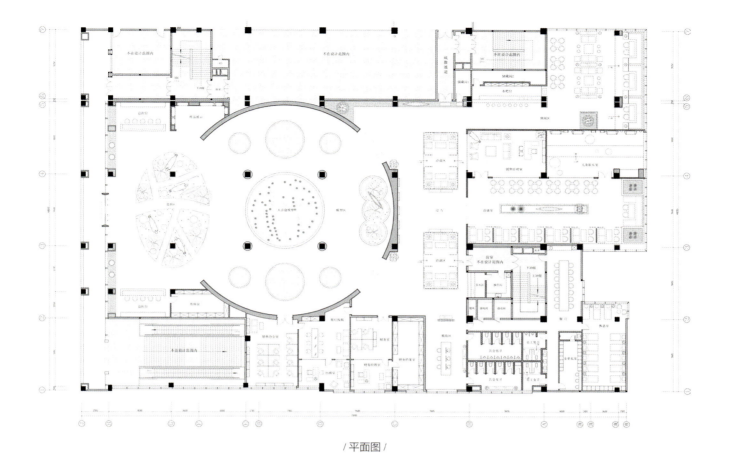

/ 平面图 /

入口由洁白的树枝、小石、缱绻云彩以及苍绿小舟引入清幽、净白的空间，诱人的阳光与白马、丛林的轻吻，让全身感官充满别有洞天的幸福之感。绿色墙体环抱而成的巨型沙盘区，青色圆管化作枝桠绵延飘浮在亮色的软膜天花之下，光影下透出丝丝的宁静于温暖。

There are white branches, stone, tempting clouds and the introduction of a green boat, also the quiet, clean and white space stay by the entrance.The attractive sun and white horse, so that the feeling of happiness is full of Journey. Surrounded by a giant wall which is made of green sand.Then it turned into a blue tube floating in the bright branches,it will give us the warm feeling.

洽谈区划成内外两个空间，外洽谈室由纯色通透方管枝条隔断围合；内洽谈区分为开放空间和半开放空间，适应不同客户的空间分流。

Negotiation area was divided into two spaces.In the outside of the negotiating room by the permeability of the pipe branches.On the inside,it is also divided into two space which is complete openning space,the other one is the half opening space.Just for varies guests to communicate.

儿童娱乐室由循序渐进的红色墙纸以及大小各异的黄、橙色圆形壁龛组合，地面铺上如流水般拥有层次分明的的苍蓝色地毯，其上散放着青绿的圆形坐垫和可爱的布偶马，构成了一个魔幻的童话空间。

Children's playroom, and by the gradual red wallpaper sizes of yellow, orange circular niche mix, the ground covered with water, such as structured as the dark green with blue carpet, the chairs were on green cushions and cute round puppet horse, form a magical fairy tale space.

厦门大韵天成设计公司办公室

重新调整过的办公室,在粗矿的水泥墙空间中通过富有想象力和张力的软性、硬性材质的相互穿插变化中营造创意的空间。入口玄关采用粗面麻质垂布穿插悬挂,并用透明的鱼丝线牵拉,在灯光衬托下形成立体感极强的玄关屏风,不经意而成的形态,使入口空间创意十足。树杈般的木龙沿卫生间墙面攀爬,宛如蓬勃生长的生命向上舒展,进一步活跃了室内空间。

Dayun Tiancheng Design Co., Ltd.

Office of the concrete wall in the coarse ore space for imaginative and tension through the soft, hard material interspersed with each other to create a creative space for change.

Entrance with rough linen cloth interspersed hanging down, and stretch silk with a transparent fish, in the light against the backdrop of the entrance to form a strong three-dimensional screen, the entrance space creative. Wooden dragon-like tree branches along the bathroom wall climbing, like life, vigorous growth, and further lively indoor space.

项目面积:180m²
项目地址:厦门九龙城
设计单位:厦门大韵天成设计有限公司
摄影:吴永长

代表作品(二)

办公环境的休闲区，利用空心水泥砖叠砌而成的吧台，悬吊的钢化清玻搁板，陶瓷咖啡杯具似乎就在空中漂浮，与悬浮式的钢构夹层相呼应，增强了空间的趣味性和想象力。运用素砼与皮革两种材质对比的办公桌以及似乎随意敲打而成凹凸质感的墙面改变了原本平淡的水泥墙……一切都使得整个办公环境显得灵动而飘逸。

Office environment, leisure areas, the use of hollow concrete blocks stacked build from the bar, the suspension of the clear glass shelf steel, ceramic coffee cup seems to be floating in the air, and enhance the interest and imagination of space. Prime concrete with the use of two materials contrast leather desk and a seemingly arbitrary changes in beating the original wall made of plain concrete walls …… All this makes the whole office seemed Smart and elegant.

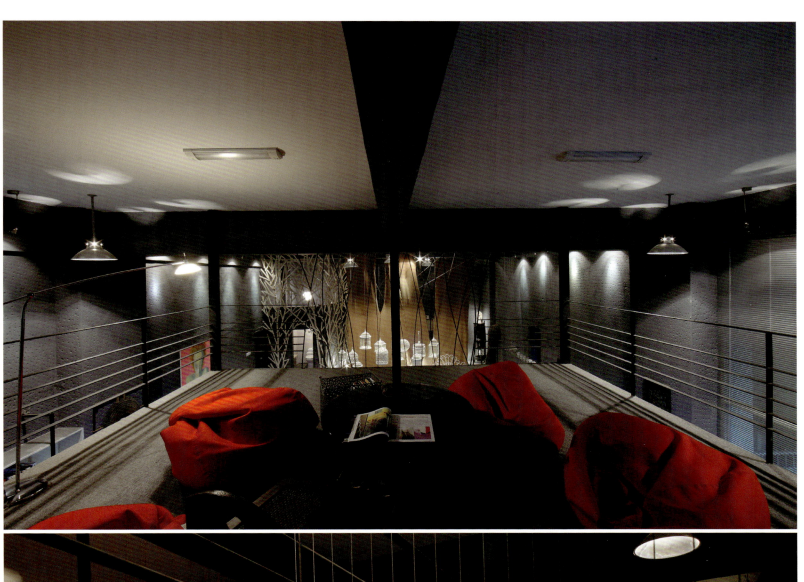

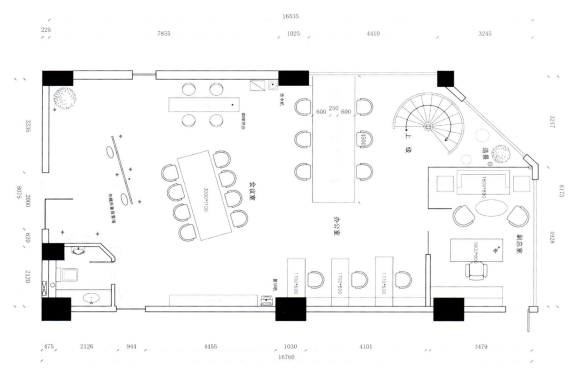

/ 平面图 /

李泷 \ Leo

- 宽品设计顾问有限公司_设计总监
- *Kuanpin Design Consultants Ltd._Creative Director*

Branch of China Institute of Interior Desi
IAI Union Pacific Architects and interior
Kuanpin Design Consultants Ltd   Desig

Major works

- Beijing Landscape Garden Club Culture

  Best Science Group (China) Headquarters
  Chong Guan Group (Hong Kong) headqu
  Jinan Yu Jingshan villa sales offices
  Saudi Aramco China Office
  Blue Stream international sales offices. m
  Beijing million city mansion model room
  Guanzhi Shan Hot Springs Resort
  Guanyinshan International Business Park

厦门室内设计在国内整体的设计环境中还处于初级水准，无论是政府主管部门的大环境还是从业者自身的专业素质以及对于行业本质的思考都有待提升。放到世界的大背景中，中国的室内设计存在同样的问题。

贵司主要从事哪方面的设计？最新的作品是什么？

- 宽品一直致力于为优质客户提供专业室内设计，如精品酒店、办公空间、会所、房地产项目及家居空间等等。
- 最新的作品包括位于鼓浪屿的那宅精品酒店以及北京的山水会所。

您的设计灵感来源于什么？

- 书籍、旅行、电影以及对于优质空间的体验。

您从事设计工作多久了？您的设计理念是什么？

- 十年。
- 宽品设计一直坚持简约而优雅的设计风格，追求作品的原创表达，努力为受众传达时尚设计理念，力求在作品中呈现设计美学与人文关怀的融合。

在设计过程中，您最注重哪方面的工作？对于多种设计元素有没有什么偏好？

- 方案构思、材质搭配以及施工过程管理。
- 宽品的作品始终贯穿着属于东方设计的思维，以及在设计中融入戏剧性和情境的营造，力图在内敛沉稳的空间塑造深层次的感染力。

从前期接触项目、出概念、做方案到完成整个项目，觉得最大的困难是什么？

- 设计理念的落实。

欣赏的或者对您影响最大的人是谁？抑或某种风格、思潮、理念？

- 贝聿铭（影响最大的设计师）；陈瑞宪（最欣赏的设计师）。

怎么看待厦门设计？以及中国设计？

- 厦门室内设计在国内整体的设计环境中还处于初级水准，无论是政府主管部门的大环境还是从业者自身的专业素质以及对于行业本质的思考都有待提升。放到世界的大背景中，中国的室内设计存在同样的问题。

未来的公司会如何发展？您本人呢？

- 公司的专业化建设是未来的工作重点，也会更加努力在设计美学的引导传播以及优质生活的理想构筑。作为宽品的一份子，同时作为设计业的从业者，在完善自身的素养同时，愿致力于促进设计的发展并实际应用于普世民生。

厦门泛华盛世

以低调诠释奢华，以简约呈现生活，融合时尚与典雅，运用不同材质的对比，简洁优雅的语汇，塑造独特的沉静气质。

整体空间的低限度色彩铺陈沉稳高雅的居家氛围，浅灰色系的丰富色阶与形体、家具、饰品等米色系色块相互辉映，缔造精致华丽的奢华视觉。

点线面的巧妙结合把餐桌、书桌及电视台连为一体，空间气势一气呵成！耐人寻味的细腻收边细部，洋溢时尚气息的陈设艺品，明朗干净的空间融合暖暖柔光，完整呈现精致生活的美学感受。

Xiamen Fanhua Shengshi

Interpretation of works of a low-key luxury to simple presentation of life, the integration of fashion and elegance, the use of different contrast materials, simple and elegant vocabulary, create a unique calm temperament.

Whole space is calm and elegant surroundings are light gray color gradation and rich body, furniture, jewelry, and other beige color embraced the luxury to create gorgeous visual refinement.

Point Line Plane's unique combination of the table, desk and television stations as a single entity, space momentum go! Revenue side of intriguing detail and delicate, filled with fashionable furnishings art, bright clean space integration of the warm soft, full of life presents exquisite aesthetic experience.

设计师：李泷
设计单位：宽品设计顾问有限公司

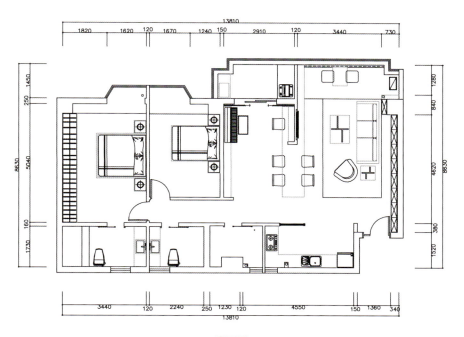

/ 平面图 /

观音山商务中心

以视觉化的具象表达，呼应企业及其产品的内在精神。以简约、素净的设计手法，塑造富含东方韵味的空间气质。

洁白、细腻的整体色调中，融合层次丰富的黑灰线体，以色彩、造型、肌理的和谐对话营造空间独特质感。从纺织产品的经纬交织中提炼线性元素贯穿于整体设计，并成为企业精神的视觉体现，为企业发展提供丰富的文化延展性。

开放、通透的设计理念，使空间的使用者能充分享受自然阳光和优美海景，并使空间更添魅力！

Guanyin Mountain Business Center

To visualize the concrete expression of the internal echoes the spirit of enterprise and its products.To simple, sober design techniques, create rich oriental flavor space temperament.

White, delicate colors, the integration of rich layers of black and gray line body, to color, shape to create a space unique texture. Extracted from the linear elements woven throughout the overall design, and become the visual embodiment of entrepreneurship, enterprise development to provide a rich cultural scalability.

Open, transparent design concept, so that users can fully enjoy the space of natural light and beautiful sea views, and to space gift to charm!

设计师：李泷
设计单位：宽品设计顾问有限公司

李学文 \ Awen

- 文阳空间设计有限公司_董事长
- Art Supervisor, Xiamen Oriental Superior Furniture Co. Ltd. President, AWEN ART SPACE

个人经历
- 国际建筑装饰协会注册_高级室内设计师
  中国建筑学会室内设计分会_会员
  中国艺术设计联盟_会员
  文阳空间设计有限公司_董事长

获奖荣誉
- 2010年10月 荣获2010亚太室内设计双年大奖赛"最佳设计方案大奖提名"
  2010年10月 荣获2010亚太室内设计双年大奖赛"亚太设计新人奖"
  2009年9月 荣获（中国风）——2009亚太室内设计精英邀请赛"2009亚太十大新锐设计师"
  2009年7月 作品入编《美国（室内设计）杂志》TOP10人气榜第二名
  2008年10月 代表作品《东方至尊国际家居馆》入选第十五届亚太区室内设计大奖商业作品选
  2008年4月 荣获2008中国（上海）国际建筑及室内设计节金外滩入围奖

Experience
- International Building Decoration Association_Senior Interior Designer
  Architectural Society of China (Interior Design Branch)_Member
  Chinese Union of Art and Design_Member
  Art Supervisor, Xiamen Oriental Superior Furniture Co. Ltd. President, AWEN ART SPACE

Awards
- Asia Pacific Interior Design Biennial 2010, Grand Prix "Best Design Program Nominative Award"
  Asia Pacific Interior Design Biennial 2010, Grand Prix, " Asia Pacific Prominent New Design Award"
  2009 Asia Pacific Interior Design Invitational " New Prominent Design Award "
  "United States (Interior Design) magazine" TOP10 Scoreboard second work
  Representative works, "International Home Gallery Extreme Orient," was selected Fifteenth Asia Pacific Interior Design Awards Commercial Selections
  2008 China (Shanghai) International Building and Interior Design Festival Finalist Golden Bund

如果可以，我希望我与我们的设计团队能影响着厦门设计市场的发展，这是我们的目标。坚持走精品路线，来提高我们的品质。因为我觉得现在厦门设计师对设计与生活有许多不一样的看法，整体还是处于初、中级发展阶段，是一种起伏不定的状态。在以后的发展路线上看是慢慢地向精品设计发展

请问您最近主要设计的项目是什么？公司主要的设计业务是什么？

- 我们现在的设计团队的设计风格是多样化的，每个人的设计思想也不同，我觉得我们是一个互补型的团体，这样有助于我们设计项目类型的多元化与设计分工的细致化。但是我们主要设计业务还是酒店与会所空间，很高兴的是近期我们主要的设计项目也是趋向酒店与会所的空间设计。

您在从事设计的过程中，您的设计灵感来源于什么？

- 我想每个设计师都有自己的设计欣赏与艺术鉴赏，我喜欢在生活细节与自然感受中寻找灵感，这样能吸取许多设计元素。在与业主的沟通中来了解业主的喜好与兴趣，并找出适合业主的风格定位与灵感的相互运用。

设计是一件很辛苦的事情，您在设计过程中遇到"思维瓶颈"的时候一般怎么办？

- 在这个时候我应该会主动与我们的团队或朋友进行自我想法的交流，这种相互间的思想看法是解决"思维瓶颈"最直接的方式。之前我说过我们团队的风格是不一样的，在许多看法上面是互补型的，所以相互的交流可以摒弃许多相互交义的想法。

鼓励或鼓舞自己从事设计工作的人或信念？

- 我会建议他给自己一个独立思考的空间，并对生活细致品位，了解不同阶层的生活喜好。所谓"细节决定成败"，设计也是对细节的一种不一样的体现方式，在细节中能激发人的设计灵魂。

有没有想过停下来，暂时不做设计，进学校学习或出国游历，然后再继续设计工作？

- 没有。个人觉得从事设计工作的过程中就是对自己生活见识的一种积累，我很享受这种设计过程，每次的方案完成或工程竣工都是一种收获与感悟，能培养个人的生活品质与生活图像。

如果不做设计，最大的愿望是什么？

- 空间的塑造是一种自我表达的理性思考，我享受这种创作过程，喜欢去了解作品的精华，如果真的不做设计的话我应该会去开阔设计的视野，去一些优秀设计作品结晶的地方，去体会艺术殿堂。

您对厦门的设计市场有何看法？

- 如果可以，我希望我与我们的设计团队能影响着厦门设计市场的发展，这是我们的目标。坚持走精品路线，来提高我们的品质。因为我觉得现在厦门设计师对设计与生活有许多不一样的看法，整体还是处于初、中级发展阶段，是一种起伏不定的状态。在以后的发展路线上看是慢慢地向精品设计发展了。

最喜欢的建筑师或室内设计师是谁？喜欢他什么？

- 现在的许多建筑师或室内设计师都是我们的导师，给我们许多指引，不过我不喜欢沉迷哪个设计师，这样也很容易影响个人的设计思维。要说喜欢的话我比较喜欢贝聿铭。我欣赏他的敬业和做事专注的态度，严谨和永远追求完美的设计风格。在他的身上能感受到那种对设计的执着。

厦门新东方至尊家居馆

该商场地处闹市区的地下一层，空间无自然采光，再加上家居商场的特性，产品展区光线较昏暗。为了使客人心情从地上到地下的豁然开朗，设计师将光作为了设计思想的全部。把家居商场的公共空间营造出超亮而清爽的时尚氛围。高亮度空间的灯光布局，必然会破坏空间的完整性。可设计师的高明在于全部运用了隐藏含蓄的打光形式，既体现了空间体面完整性的同时又做到了投光的朦胧美。整个空间神秘而清澈，前卫而高雅，超脱了传统的空间理念，让人置身世外极地，又如极地的圣光，让人心旷神怡！

New Oriental Extreme Home Gallery

The mall is located in the basement of downtown, the space without natural light, combined with the characteristics of shopping, product exhibition warm dark than light. In order to give guests the feeling from the floor to the ground suddenly. Designers will be light as the design concept of all. The mall's public space to create a bright and refreshing style atmosphere. Spatial layout of high-brightness light will inevitably undermine the integrity of the space. Clever designer can use the hidden implication is that all the lighting in the form, not only embodies the integrity of the space while decent did cast light on the hazy beauty. Mystery and clear the entire space, avant-garde and elegant, detached the traditional concept of space, people exposure to the world outside the polar regions, and if the polar of the Light, people relaxed and happy！

设计师：李学文
设计单位：文阳空间设计有限公司

厦门庄式家居新馆

本家居馆为厦门家居的名牌老店，新馆的风格定位为时尚、简约、大气。店中的家居产品风格多为时尚、新颖、高贵。设计师在充分理解业主的意图后，将现代简约的手法融入到尊贵高雅的空间氛围中。运用大手笔的块面表现、水晶线条和黑钛天花的相交互映、光源的散落分布和钢琴流水的巧妙结合，映照出商场大堂中心高贵和典雅的氛围。

Zhuangshi Home Gallery

The home of the famous museum home for the old shop in Xiamen, a new positioning for the fashion museum style, simplicity, the atmosphere. Style of home products store, mostly fashion, new, noble. Designers to fully understand the intent of the owners, will be modern and simple way into the noble and elegant atmosphere of space. Generous use of the performance of the block surface, crystal and black titanium line interactive mapping phase of smallpox, the scattered light distribution and flow of the unique combination of piano, reflects the noble and elegant shopping center lobby of the atmospheric environment.

设计师：李学文
设计单位：文阳空间设计有限公司

将楼梯作为中心表现的主体,使顾客自然而欣悦的迈向二楼,巧妙地解决了商场人流导线的难点问题。

The performance of the main body of the stairs as the center, so that customers move towards natural and shown a second floor, cleverly solved the difficult problem of wire shopping crowd.

通道及水吧等空间，都与大堂天花的斜线灯带相互融通，空间处理上则打破了传统家具商场的区域划分理念，统一而简约的表现手法和大堂的整体气质相衬托，更加突出了整个商场的空间氛围：简约、时尚、大气且高雅，使人流连忘返。

Channels such as space and water bar, the lobby ceiling with the lights with another slash facilities, spatial processing is broken on the regional division of the traditional furniture store concept, the performance of a unified and simple approach and set off the lobby of the overall temperament, more prominent Space throughout the mall atmosphere: simple, stylish, and elegant atmosphere, people forget.

- 厦门刘腾华室内设计顾问有限公司_创意总监
- *Xiamen Liu Tenghua Interior Design Consultants Ltd. _Creative Director*

个人经历

- 厦门刘腾华室内设计顾问有限公司_创意总监
  毕业于福建师范大学美术学院

Experience

- Xiamen Liu Tenghua Interior Design Consultants Ltd._Creative Director
  Graduated from Fujian Normal University Institute of Fine Arts

厦门设计市场的发展归结于设计师对设计的态度，很自豪我们厦门这么多优秀的同行都在为厦门设计走向国际而不屑的坚持着，同时我们需要更多的学习、交流、沟通、探讨、分享、鼓励、喜悦……

请问您最近主要设计的项目是什么？公司主要的设计业务是什么？

- 主要是家居设计、商业店面等，公司主要的设计业务之前基本都是生活家居类案子，近来开始做一些商业类案子，还初涉部分设计型小酒店。

您在从事设计的过程中，您的设计灵感来源于什么？

- 不喜欢华而不实的东西，向往自然不需刻意雕琢的一切，想法归结于我的成长我的经历和游历，我喜欢去观察和研究生活中一些我认为美的事物，从中去体会它们的精神，不定时出去游历也是我一部分想法的来源方式……

设计是一件很辛苦的事情，您在设计过程中遇到"思维瓶颈"的时候一般怎么办？

- 我的生活方式都是比较缓慢、比较随意，跟着心情在走，从另一种方式来说我也比较懒（哈哈），所以一般不会把自己整得累累很辛苦，要懂得自我调节，我崇尚边设计边玩，找个我愿意呆着的地方睡懒的躲起来，或者一个人没目的随意出去走走，出走是一种身心完全放飞的感觉，去陌生的地方感受不同的文化和乡土人情，我每年总要有几次去享受那样的生活状态，有时确实觉得设计是非人的工作，但痛并快乐着，喜欢这种感觉！

鼓励或鼓舞自己从事设计工作的人或信念？

- 因为喜欢，所以坚持，因为挑剔，所以凡事都要做到极致……

有没有想过停下来，暂时不做设计，进学校学习或出国游历，然后再继续设计工作？

- 我现在的状况都是边设计边学习边积累，因为还是有太多的不足，设计是一项学到老的学科，需要我们有一颗不断进取的心和一颗懂得欣赏美的心，其要用心去感受和体会……

如果不做设计，最大的愿望是什么？

- 没什么大愿望，还是做自己喜欢做的事情，去我想去的地方，随缘，无拘无束，一切随当下的心境去飞。

您对厦门的设计市场有何看法？

- 厦门设计市场的发展归结于设计师对设计的态度，很自豪我们厦门这么多优秀的同行都在为厦门设计走向国际而不屑的坚持着，同时我们需要更多的学习、交流、沟通、探讨、分享、鼓励、喜悦……

最喜欢的建筑师或室内设计师是谁？喜欢他什么？

- 很长一段时间喜欢苔林徽因、梁思成；我会不定期喜欢一些人，喜欢他们的思想！

香槟城

本案的业主是一对年轻的新婚夫妻,由于居住人口单纯,客、餐厅以及书房是两人使用密度最高的区域,因此全案在公私场域的配比上保留了最宽敞的公共场域,给予这对新婚夫妻自由无拘的生活感受。

在起居生活区,拆除厚重的砖墙,借助白色木制叶脉的格栅界定出客厅、书房、餐厅的空间关系,融洽而有趣味。

Champaign City

This case the owners are a pair of young newlyweds, the resident population alone, passenger, restaurants, and two of the study is to use the highest density regions, so the whole case, the ratio of public-private field to keep the most spacious public spheres, Give to newly married couples feel free life no arrests.

Living life in the area, removal of thick brick walls, with white wooden grille veins define the living room, study, dining room relationship, harmony and have fun.

设计师:刘腾华
设计单位:厦门刘腾华室内设计顾问有限公司
摄影:刘腾飞

代表作品(一)

/ 平面图 /

深灰与浅灰的墙面带来的宁静,白色的烤漆钢制书架、鞋柜、格栅与黑色的家具的对比,粗犷纹理的木地板在柔和的灯光烘托下布满整个空间,静谧而不失活力。

Dark gray and light gray walls bring quiet, white painted steel shelves, shoe racks, grille and black furniture contrast, the rough texture of the wood floor under the contrast in the soft light filled the entire space, quiet Without losing energy.

郭氏家居

我们想展现一个细腻耐看的空间特质，为居住者勾勒一个以美学为基础的生活容器，提升居住者对美感的讲究与认知，而这种提炼的过程，必须是讲究生活，享受生活所带来的情趣，业主多半需要具备相当的成熟与经济力。

本案屋主为一企业主，此户提供屋主的起居生活，三户打通的超大尺度却只需承载三人的生活作息，我们总在寻找那份符合屋主心灵归属感的空间，透过每一细节的讲究，让生活的况味能够展现完全的从容、舒展与惬意，这也是本案的核心。

Mr. guo Room Design

We want to show the spatial characteristics of a fine engaging for residents outlining the basis of an aesthetic life of the container to enhance the occupants of the aesthetic and cognitive stress, and this refining process, must pay attention to life, enjoy life brought To the taste, most owners need to have considerable maturity and economic power.

This case the owner is a business owner, provide owner households living in this life, three to get through the very large scale is just carrying the three men lifestyle, we are always looking for a sense of belonging mind share of the space meet the owner, through the Pay attention to every detail, so that conditions of life to show the full taste of calm, stretch and comfortable, which is the core of the case.

设计师：刘腾华
设计单位：厦门刘腾华室内设计顾问有限公司
摄影：刘腾飞

代表作品（二）

/ 平面图 /

邵力中 \ Shao Lizhong

- 厦门世纬设计事务有限公司董事_设计总监
- Xiamen Shi Wei Co. Ltd. _ Creative Director

邵力中
- 同济大学建筑学学士_建筑设计师
  高级室内设计师
  厦门世纬设计事务有限公司创始人_设计总监
  上海同鸿规划建筑设计有限公司董事_设计总监
  厦门京闽中心酒店_艺术顾问
  1989-2009中国室内设计二十年杰出设计师（CIID）
  2009中国室内设计年度封面人物
  (CIID/INTERIOR DESIGN)

Shao Lizhong
- Tongji University, Bachelor of Architecture_architect
  Senior Interior Designer
  Xiamen Shi Wei, founder of Design Director Design Services
  Planning and Design Shanghai with Hong Design Director, Director
  Xiamen Central Hotel Art Consultants
  1989-2009 China's two decades of outstanding interior design designer (CIID)
  2009 China Interior Design Annual Cover (CIID / INTERIOR DESIGN)

姜辉 \ Jiang Hui

- 厦门世纬设计事务有限公司董事_设计总监
- Xiamen Shi Wei Co. Ltd. _ Creative Director

姜辉
- 大连理工大学建筑学学士_建筑设计师
  高级室内设计师
  厦门世纬设计事务有限公司董事_设计总监
  2001-2002福建省优秀室内设计师
  2010全国资深室内设计师

Jiang Hui
- Dalian University of Technology Bachelor of Architecture
  Architect Senior Interior Designer
  Xiamen Shi Wei, Director Design Director Design Services
  2001-2002 Fujian Excellent interior designers
  2010 National Senior Interior Designer

杨琳 \ Yang Lin

- 厦门世纬设计事务有限公司董事_总经理
- Xiamen Shi Wei Co. Ltd. _ General Manager

杨琳
- 同济大学建筑学学士_建筑设计师
  高级室内设计师
  清华大学EDP房地产开发与金融运作总裁班
  厦门世纬设计事务有限公司董事_总经理
  1989-2009中国室内设计二十年优秀设计师（CIID）

Yang Lin
- Tongji University, Bachelor of Architecture_architect
  Senior Interior Designer
  Tsinghua University, EDP, president of real estate development and finance classes
  Xiamen Shi Wei, Managing Director of Design Services
  1989-2009 China's two decades of outstanding designers Interior Design (CIID)

> 我们不建议将建筑的室内外分割开讨论，这应该是一个完整的创作过程。项目完成之后所呈现的，不仅仅是风格和美感，同时还有其不可或缺的社会属性。如何取得一个项目在社会、经济、文化和美学等各个范畴的综合平衡，正是我们的课题。

你们从事设计工作多久了？贵司主要从事哪方面的设计？

- 我们从1992年就开始各自从事设计工作，在1998年注册了公司。
- 2005年之前，我们接触的项目非常广，基本上所有的室内项目形式都经历了，目的是积累经验。2005年之后，我们会精选一些项目，比如大型的综合商业体、高端酒店等。因为在这些项目类型上我们做了很多研究，有很多心得和经验。

可以分享一下你们的心得和经验吗？

- 高端酒店项目，我们要求定位准确，并在投入运营后有良好的业绩反应。我们很注重市场分析，与酒店管理集团互动较多。我们接触的项目主要是私人或集团投资，国际品牌酒店和四星以下的酒店并不适合这种资本模式，所以我们建议业主要对相关业态进行全面的了解。
- 再就是大型的商业综合体，比如shoppingmall、步行街。我们对这种商业项目有比较深的认识，2000年我们设计了福建省第一个真正意义上的mall，至今闽南地区比较大的商业体都是我们负责的。在每一个项目中我们会多方面介入，我们三个人都是学建筑的，所以擅长从概念、模式、结构上做整体的考虑。甚至于资本运作、商业运营方面，我们的模式和运作都很成熟，概念是我们的优势，经验是我们的强项。

最近比较满意的作品是什么？

- 九间房七方院，它是一个会所，建在一片60亩的原始生态果林中。我们负责该项目从景观到建筑到室内空间的全面设计，它体现了我们对闽南古建筑的研究和体会。作为建筑学专业的毕业生，虽然我们的工作重点在室内空间，但对于建筑的热爱是丝毫没有减退的，这个项目的完成也算是了却我们多年的一个心愿。
- 我们对闽南当地的古民居很感兴趣，也一直在研究。闽南话就是一种活化石，遵循了很多传统，和他们的建筑是息息相关的。闽南的古民居是以整体村落的形式存在的，依旧跟古代保持一致，在制式上遵循传统规范。我们沿袭原来建筑的精神，重新组织形式，变为现代人可以居住的环境。

对现代中式风格有什么认识？

- 回归传统是一种本能的需求，近年来人们越来越渴望回归自己的内心。现代中式风格，大多数体现在装饰上，比如运用很多中式家具和装饰元素。但是现代建筑和传统的建筑完全不一样，这种装饰只是很表面的一种方式。建筑是一种大的感觉，一种秩序，一种对原则的把握。中式建筑体现了一种轴线的礼制，是空间上的秩序、主次、轴线关系等。
- 传统的制式是精神，也是一种文化，制式蕴含了古人对各种关系的思考，轴线、小路、厅、房、功能分区有非常多的讲究。比如：我们在九间房七方院项目中重新结构古厝的制式，重组轴线，调整局部，重新组织。形式、材料是现代的，大片的玻璃、贴砖，我们希望在空间上来体现这种回归，而不是简单的装饰和个别的元素。

在设计过程中，您最注重哪方面的工作？设计理念是什么？

- 我们从建筑阶段或建筑改造阶段介入项目，包括景观、室内到最后的完成，甚至开业过程。我们追求的不是外在的形式，比如扎哈·哈迪德的建筑非常酷，但是造价很高，投入产出比难以估算，这对业主是不负责任的。我们不建议我们的业主做这样的形式。
- 创新是困难的，绝大多数的创新是一种表皮处理。人类文化几千年，对于空间，已经没有太多创新的可能，更重要的是如何使用空间，创造价值。所以我们会对业主进行"洗脑"，怎么设计、经营、维护，我们会全程为他出谋划策。

你们怎样和业主沟通并达到"洗脑"效果？

- 我们的客户比较稳定，彼此忠诚度很高。我们的合作目标是双赢。帮助业主减少投入，帮他实现对建筑、对空间的理想，达到他期望的经营效果。我们的商业综合解决能力、策划能力很强。这些问题解决了，业主就会非常信任我们，设计风格上给我们比较大的自由，图纸的贯彻性也比较好，解决了施工问题。

你们怎么进行管理？

- 我们的项目规模大，社会责任也更强，所以我们的设计师并不是能够速成的。每个年轻人大概需要培养3年才能够进入项目。我们看重的是每个人在团队中的位置，在自己的位置上能够尽职尽责。

未来的公司会如何发展？你们个人呢？

- 我们会选择项目、选择业主，也一直在控制公司的规模。
- 我们希望更好地生活，注重生活品质，提倡把握快乐、而有质感的人生。我们希望设计工作只是我们生活的一部分，希望团队的每个人都会"玩"，生活的很快乐。

九间房七方院

"九间房七方院"是我们对于中国传统建筑的房院体系的一次现代化探索，也是在当代社会全球化思潮中对人们心灵深处中国情怀的一次挖掘。建筑的轴线在这里被隐含了，院落与房屋彼此有力度地相互穿插，虚实的对比更多地被转移到墙体和屋面的关系上，视线在穿越了转折多变的院墙时，也穿越了透明的屋顶，投向天空和更远的山体。
中国建筑师的灵感总是来源于内心，而又更多地表现在房与院的微妙关系上。

Rooms & Courtyards

Work is our system of traditional Chinese courtyard building a modern exploration of globalization is also a trend in contemporary society to people in the feelings of a soul mining in China. Axis of the building is hidden here, the courtyard and housing efforts with each other interspersed with each other, the actual situation of the comparison is more often transferred to the relationship between the wall and the roof line of sight through a turning point in changing the walls, they Through a transparent roof, toward the sky and the mountains beyond.

项目面积：265 m²
竣工时间：2008年4月
主创设计：邵力中／杨琳
建筑设计：林铮顗
摄影：申强

代表作品（一）

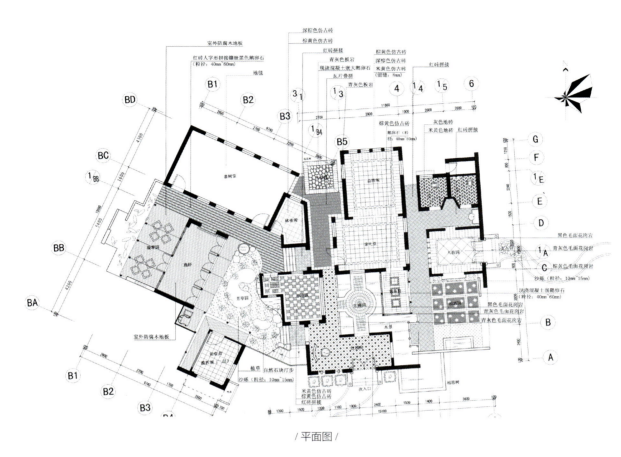

/ 平面图 /

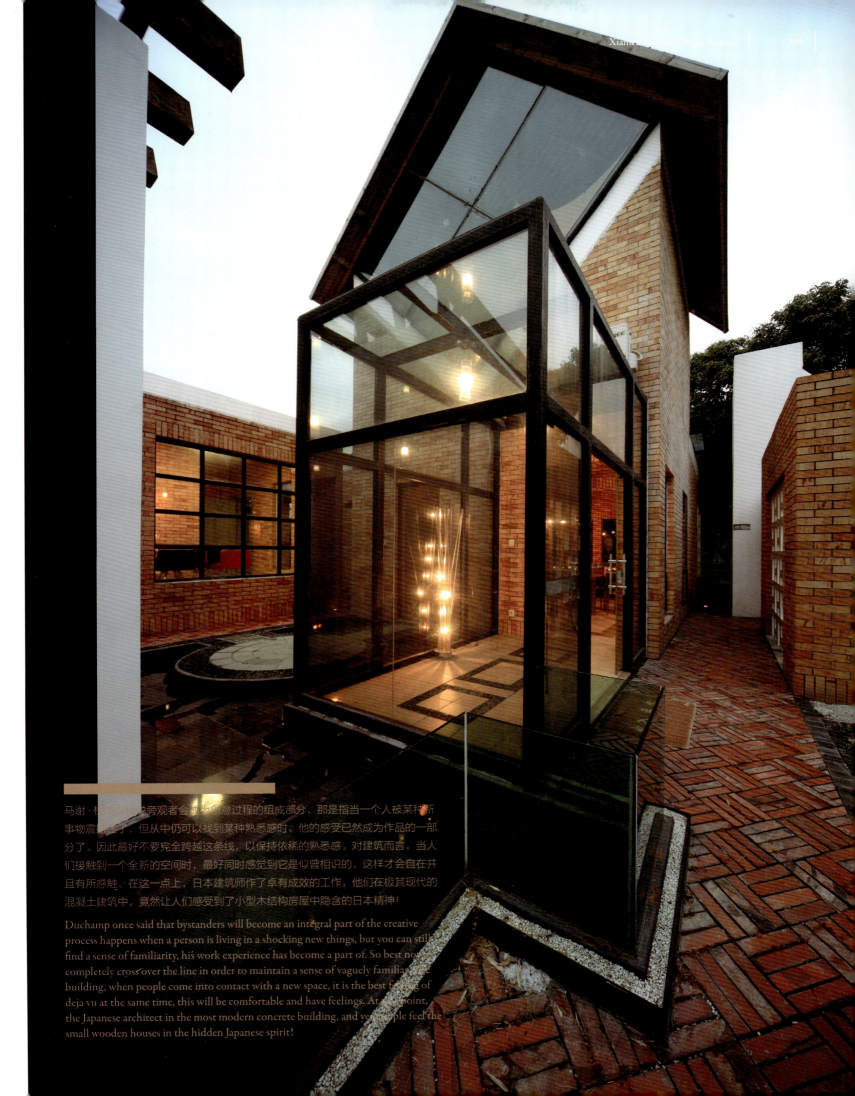

马谢·杜尚曾经说旁观者会成为创意过程的组成部分，那是指当一个人被某种新事物震撼住了，但从中仍可以找到某种熟悉感时，他的感受已然成为作品的一部分了。因此最好不要完全跨越这条线，以保持依稀的熟悉感。对建筑而言，当人们接触到一个全新的空间时，最好同时感觉到它是似曾相识的，这样才会自在并且有所感触。·在这一点上，日本建筑师作了卓有成效的工作，他们在极其现代的混凝土建筑中，竟然让人们感受到了小型木结构房屋中隐含的日本精神！

Duchamp once said that bystanders will become an integral part of the creative process happens when a person is living in a shocking new things, but you can still find a sense of familiarity, his work experience has become a part of. So best not completely cross over the line in order to maintain a sense of vaguely familiar. For building, when people come into contact with a new space, it is the best feeling of deja vu at the same time, this will be comfortable and have feelings. At this point, the Japanese architect in the most modern concrete building, and yet people feel the small wooden houses in the hidden Japanese spirit!

中国其实也不乏这类的探索，我们的老师冯纪忠先生早在上世纪80年代初就在上海的近郊设计了松江方塔园，那直棱棱的坡顶现代感十足，但又实实在在地表现着中国传统，成为那个年代屈指可数的优秀作品。当然在近年来我们似乎也在尝试重拾传统，比如"鸟巢"中包裹的中国式的红鞋，比如苏州博物馆的新式的粉墙黛瓦，比如运河岸上的房子……它们的共同点是：时尚。作为当代的建筑作品，它必须是时尚的，这种"时尚"指的是它所采用的材料、所选用的建造手法，包括所提供的功能，都应该是现代的、科学的。

In fact, China has no shortage of such exploration, our teacher Mr. Feng Jizhong back in the early 80s of last century on the outskirts of Shanghai Songjiang Tayuan designed as a handful of outstanding works of that era. In recent years, it seems that we are trying to revive traditional, such as "Bird's Nest"in the package Chinese-style red shoes, the new Suzhou Museum, tile, canal, house … … they have in common are: fashion. As a contemporary architectural works, it must be stylish, this "fashion"refers to the materials it uses, the choice of construction techniques, including the functionality provided, should be modern and scientific.

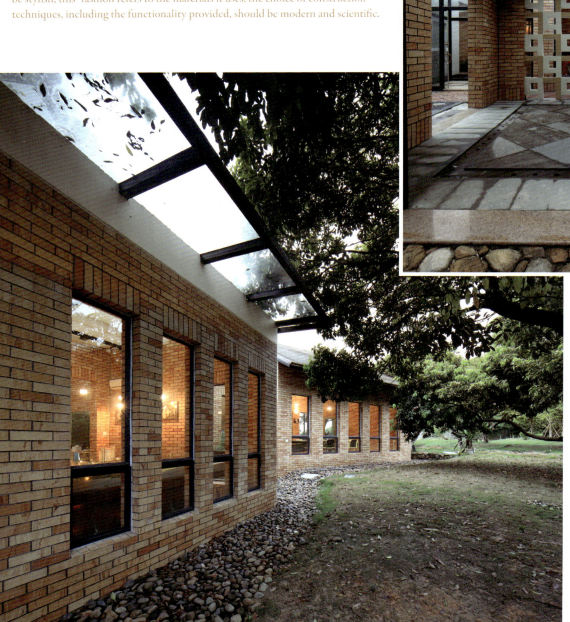

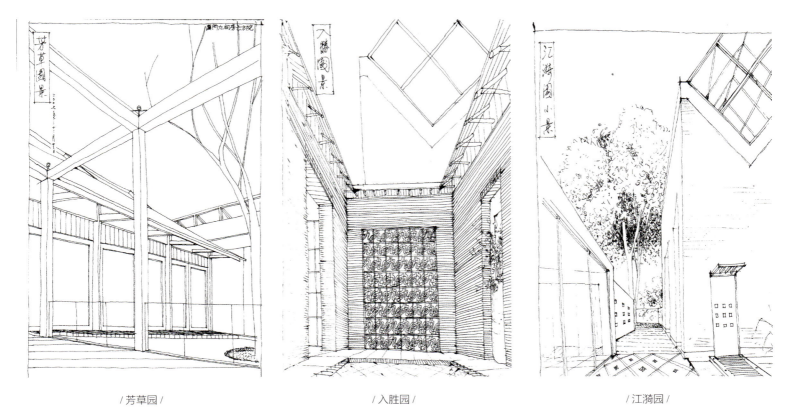

/ 芳草园 /　　　　　　　　　/ 入胜园 /　　　　　　　　　/ 江漪园 /

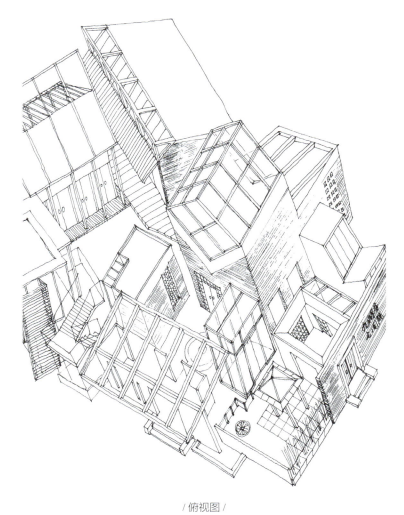

/ 俯视图 /

中国在历经了几十年对文化传统的颠覆后，迎来了经济的高速发展和物质的极大繁荣，在经历了物欲膨胀和享乐的快感之后，有些人开始思考，他们开始反对那些没有人情味的简单的所谓现代建筑，他们渴望得到空间的人文关怀。也许隐身在厦门竹坝的那片密林中的"九间房七方院"，能在这方面作一个探索，引起闽南地区现代人的些许中国式感应。

China has experienced after decades of cultural traditions, ushered in rapid economic development and material of great prosperity. After the boom, some people start thinking that they crave space, humane care. Perhaps hidden in Xiamen, the plot of the bamboo forests in the dam of this work can be an exploration in this area, causing a bit of modern China's southern region of induction.

厦门瑞景商业广场

瑞景商业广场是集休闲购物、餐饮娱乐、酒店公寓、商务办公于一体的城市商业综合体。整体规划占地5.7万㎡，总建筑面积为8.7万㎡，其中商业面积6万㎡，由2万㎡"好又多"大卖场、4万㎡回廊步行街和主题商业组成；酒店式公寓、商务办公楼1.7万㎡，停车位430个。

Commercial center of Xiamen Ruijing

Business Centre is a leisure shopping, dining and entertainment, hotel apartments, commercial office in one city commercial complex. Overall planning area of 57,000 sqm, total construction area of 87,000 sqm, of which 60,000 sqm of commercial space by 20,000 sqm, "Trust-Mart," hypermarkets, 40,000 sqm commercial corridors composed of Walking Street and themes; Apartments, commercial office buildings 17,000 sqm, 430 parking spaces.

建筑面积：8万㎡
主创设计：姜辉／邵力中
景观设计：吴晓燕
室内设计：罗丽明／叶文博／叶敏娟
摄影：申强

代表作品（二）

建筑结构形式：回廊步行街、主题商业、酒店式公寓、商务办公楼。从设计构思到竣工开业，只经历了不到一年的时间。我们的设计范围涉及到景观、室外广场和所有的室内公共部分。瑞景商业广场以优秀的设计，创新的理念，注重运用节能环保技术等而被厦门市贸发局推选，作为厦门惟一的商业代表，参加商务部举办的第三批"全国商业示范社区"的评选。目前该项目已经成为厦门岛东部重要的商业综合体，并获得了2006年中国商铺百强价值品牌最佳商业地产代表作和2009年"中国土木工程詹天佑奖"金奖。

Building Structure: Walking Street corridor, subject commercial, apartments, commercial office buildings. From design conception to completion of the opening, only experienced less than a year, related to the scope of our landscape design, outdoor plaza, and all indoor public part. Ruijing commercial plaza with excellent design, innovative ideas, focus on the use of energy saving technology such as the TDC was the selection of Xiamen, Xiamen, as the only business representatives, organized by the Ministry of Commerce to participate in the third installment of the "National Business Model Community" selection. Currently the project has become an important commercial east Xiamen Island complex, and received a hundred shops in China in 2006 the value of the best brands of commercial real estate representative and in 2009, "China Civil Engineering Construction Zhantianyou Award" Gold Award.

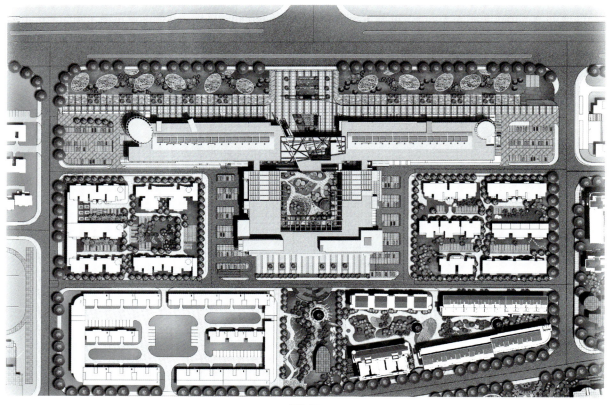

/ 平面图 /

石狮建明国际酒店

酒店选址于石狮市的市区边缘，建设初期周边还是以农田和私人建筑为主的典型的城乡结合部景象。如今，这里已成为石狮乃至泉州地区的明星区域，除了市政规划的引导外，一个五星级酒店对于周边地块的带动作用十分明显。

酒店是由一座独立罗马式大堂、一栋9层主楼、一栋10层的辅楼和一栋24层的酒店式公寓组成的城市商务娱乐酒店。我们的方案以建筑设计为起点，以我们最擅长的室内设计为主线，结合主题明确的室外建筑及景观设计，在对当地市场进行充分研究的基础上，发挥团队的优势，在成本可控范围内，用最短的时间建起了当地第一家五星级酒店。

Jian Ming International Hotel of Shishi

Hotel is located in Shishi City, the urban fringe, the early construction of the surrounding farmland and private buildings is dominated by the typical scene of urban fringe. Today, this place has become a star in Quanzhou, Shishi and even regional, in addition to municipal planning, the addition of a five-star hotels around the block for the leading role is obvious.

Hotel is an independent Roman-style lobby, a 9-storymain building, a 10-story annex building and a 24-story hotel-style apartment buildings of the city business and entertainment hotel. Our solution to architectural design as a starting point to our best at the main line of interior design, combined with subject-specific outdoor architectural and landscape design, a full study on the local market, based on the play to the team's advantage in cost manageable within the shortest possible time with the first locally built five-star hotel..

建筑面积：6.4万 m²
主创设计：邵力中／姜辉
景观设计：吴晓燕
室内设计：叶志坚／陈锻炼
陈设设计：杨琳／洛微
摄影：邵力中／姜辉

代表作品（三）

经过多年的实践和总结，我们将团队目标集中在综合性商业地产类的项目服务上，这其中包括了高星级酒店和商业广场、步行街、商业办公楼宇的综合设计服务。我们总结出作为一个成熟的酒店设计团队，它的使命在于：第一阶段，达到功能上的完整和流线的顺畅；第二阶段，获得适当的美学价值；第三阶段，营造一个酒店的空间架构的同时，满足其作为综合性盈利产品的基本条件，取得良好的内部运营效果和理想的投资回报；第四阶段，传达作为投资人、管理者和设计者的社会责任，带动消费，创造就业，提升城市形象，引导生活方式，成为城市的新名片；第五阶段，物业的升值，这考验一个酒店设计团队对于市场反应、投资控制、风格理念和前瞻性设计等多方面素质。前两个阶段目标是许多室内设计师或小型团队能够完成的，而后三个阶段目标则需要专业化团队通过实际项目的长期磨练才有可能达成。

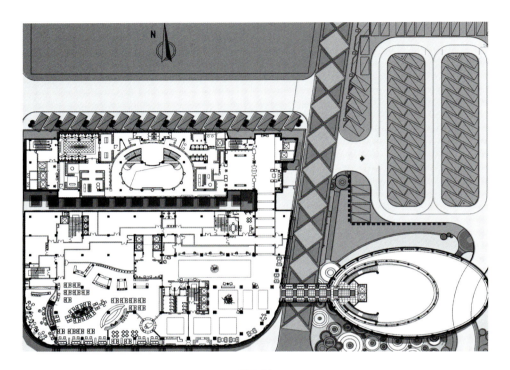

/ 平面图 /

After years of practice and lessons, we will focus on team goals of comprehensive commercial real estate projects like the service, which includes a high-star hotel and business plaza, pedestrian, commercial office buildings in the integrated design services. We conclude that as a mature hotel design team, its mission is: the first stage, to the functional integrity and smooth flow lines; the second stage, appropriate aesthetic value; the third stage, creating a hotel room structure at the same time, meet as a comprehensive basic conditions for profitable products, and achieved good results and good internal operations of investment return; the fourth phase, to convey as investors, managers and designers of social responsibility, promote consumption and create employment enhance the city image, and guide way of life, to become the city's new business cards; fifth stage, the property's appreciation, which is a hotel design team to test the market response, investment control, style, philosophy and forward-looking design, and many other qualities. The first two milestones for many interior designers or a small team can accomplish, then you need to target the three stages of professional team honed through practical projects is possible to achieve long-term.

幸运的是，在本项目上，我们遇到了开明同时有实力、有魄力的投资方，让我们的理念得以部分实现，作为在设计道路上不断探索的阶段性成果，得到了社会各界的认同。

Fortunately, in this project, we encountered an open-minded at the same time strong and enterprising investors, let us be part of the realization of the concept, as in the design of the road continue to explore the initial results obtained agree with all sectors of society .

汤建松 \ Tang Jiansong

- 香港和丰设计顾问（厦门）有限公司_创意总监
- Hong Kong Design Consultant abundance (Xiamen) Co., Ltd._Creative Director

个人经历
- 2003年，毕业于福建工艺美术学院
- 2005年，厦门喜玛拉雅装饰设计公司_设计师
- 2006年，厦门共想设计装饰工程有限公司_设计总监
- 2007年，创立香港和丰设计顾问（厦门）有限公司

代表作品
- 2010年，获香港亚太室内设计大奖赛办公组_金奖
- 2010年，获香港亚太室内设计大奖赛商店组_铜奖
- 2010年，获香港亚太室内设计大奖赛展览组_铜奖
- 2009年，获香港亚太室内设计大奖赛商店组_铜奖
- 2007年，获香港亚太室内设计大奖赛办公组_优胜奖
- 2006年，获香港亚太室内设计大奖赛商业组_铜奖

Experience
- In 2003 Graduated from the Fujian Academy of Art
- In 2005 Himalaya decoration design designer Xiamen
- In 2006 Xiamen were to design director of Design & Decoration Engineering Co., Ltd.
- In 2007 Founded the Hong Kong Design Consultant abundance (Xiamen) Co., Ltd.

Major works
- In 2010, the Hong Kong Asia Pacific Interior Design Competition Gold Medal in office
- In 2010, the Hong Kong store group of the Asia Pacific Interior Design Competition Bronze
- In 2010, the Hong Kong Asia Pacific Interior Design Competition and Exhibition The Bronze Award
- In 2009, the Hong Kong store group of the Asia Pacific Interior Design Competition Bronze
- In 2007, Asia Pacific Interior Design Competition by the Hong Kong office group prize
- In 2006, the Hong Kong business group of the Asia Pacific Interior Design Competition Bronze

> **厦**门设计行业起步比较晚，但是这几年来得到了很大的发展。厦门设计的差异比较大，风格各异，设计师之间缺乏交流。中国设计行业发展迅速，但是专注于设计的年轻设计师生存压力也比较大。没有名气，就很难接触到好的项目；为了生存，又不得不接受普通的项目。酒香也怕巷子深，所以我们现在会尽量参加比赛。

您从事设计工作多久了？能简单介绍下吗？

- 2005年，我进入喜马拉雅设计装修有限公司工作。2007年辞职，我跟方令佳合伙开设计公司。因为市场不成熟、个人经验和客户资源不足、定位不准确，公司只维持了11个月。之后我们又各自成立了自己的公司，现在发展都不错，我们也保持交流、合作。

贵司主要从事哪方面的设计？最新的作品是什么？

- 小空间为主，1000m² 以内的项目。做大的项目需要专业、庞大的团队，我目前还没有这样的控制力。最近在做一个展示空间。但是相对于功能复杂的展示空间、零售餐饮空间，我更喜欢设计功能简单的空间，比如办公室。

在设计过程中，您最注重哪方面的工作？对于多种设计元素有没有什么偏好？

- 我之前是学工业设计的，没有受过专业的教育。因此少了行业传统的限制，在造型上更加敏感自由。我在做设计的时候很少从实践、施工的角度出发，我不喜欢接受现实条件的约束，倾向于在一种理想状态下完成概念，然后再来解决实际的问题，一步步地实现方案。

如果碰到方案实践的技术难题怎么办？会更改方案吗？

- 这种情况在我从事设计工作的早期会发生。而现在，我的思路、风格都比较成熟，基本上不会遇到这种问题了。

从事设计工作的过程中，觉得最大的困难是什么？

- 最大的困难还是市场。业主是不尽相同的，第一种业主即市场上居多数的客户群体，我们必须要按照他的意愿来进行设计。这类方案不需要很高的设计水平，只需解决基本问题，外表漂亮。最后的方案往往是业主认可，但是我们自己不满意。还有一种业主是朋友，可以跟业主之间进行平等的交流，尽可能地实现最初的设计理念，但是设计费不高。

怎么进行设计管理？如何跟深化、施工部门合作？

- 施工有其非常标准的流程，从施工图纸、工地的监理到最后软装的搭配，都是严格按照顺序进行，每一步都要有专人跟踪负责。国内在这方面的标准并未被完全执行，比较随意。我们公司前期设计完后，中期会有项目经理专门跟踪项目施工，后期有专门负责软装的设计师配合业主进行软装的搭配。深化的过程可能比设计还要复杂，所以我几乎每天都会去工地。只有在现场，你才能发现一些图纸上考虑不到的细节问题，然后进行细部调整。

怎么看待厦门设计？以及中国设计？

- 厦门设计行业起步比较晚，但是这几年来得到了很大的发展。厦门设计的差异比较大，风格各异，设计师之间缺乏交流。

- 中国设计行业发展迅速，但是专注于设计的年轻设计师生存压力也比较大。没有名气，就很难接触到好的项目；为了生存，又不得不接受普通的项目。酒香也怕巷子深，所以我们现在会尽量参加比赛。

未来的公司会如何发展？您本人呢？

- 年轻的设计师比较敢为，不保守，也不会过多地考虑公司运营情况。以前我也是这样，不考虑现实因素，尽量把方案做很酷很炫，并极力实现。现在我的设计观念也发生了转变。我会考虑很多现实的条件，比如成本、工期等等。现在我追求空间的持久性，我希望我的设计能在很长一段时间都不过时，并让人们在使用的时候保持方便舒适。

东方喜意有限公司

在几何形体的交错中，我们用结构本身塑造了整个外观，凌乱的线条和起伏的轮廓，在与灯光的碰撞及融合中形成一种强烈的视觉冲击，而随性的玻璃和夸张的门洞，更加增添了这种震撼的力度。从而把整个结构朴素的质感和不羁的气质，无限放大。这些本身各异的元素，交织成一个抽象的实木装置……

Dongfang Xiyi Ltd.

In the staggered geometry, we use the structure itself shaped the appearance, messy lines and undulating contours, in the light of the collision and fusion with the formation of a strong visual impact, and with the nature of the glass and exaggerated openings, Further added that the intensity of shock. Thus the entire structure of simple textures and uninhibited temperament, hugely magnified. These different elements of their own, are woven into an abstract wood device ……

项目地点：厦门市软件园二期观日路84栋4F
建筑面积：700 m²
主创设计：汤建松

代表作品（一）

Xiamen Interior Design Annual

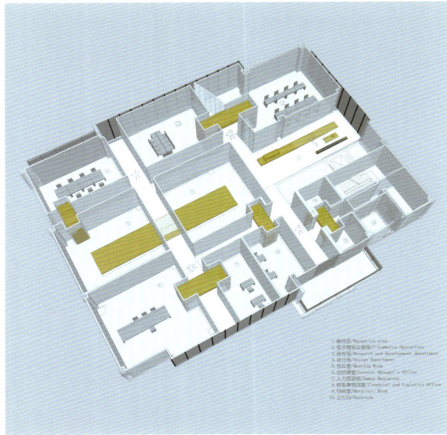

1. 接待区/Reception area
2. 电子商务运营部/E-Commerce Operations
3. 研发部/Research and development department
4. 设计部/Design Department
5. 会议室/Meeting Room
6. 总经理室/General Manager's Office
7. 人力资源部/Human Resources
8. 财务暨物流室/Financial and Logistics Office
9. 材料室/Materials Room
10. 卫生间/Bathroom

/ 轴视图 /

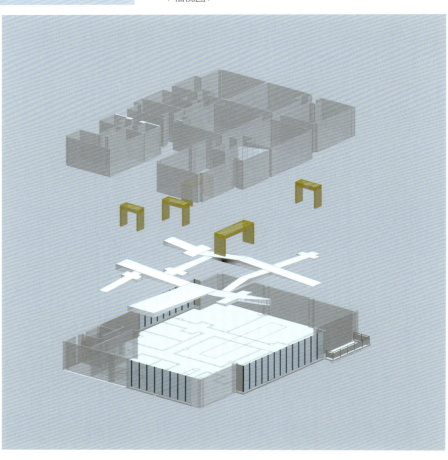

/ 轴视图 /

香港和丰设计公司办公室

古香古色的材质穿梭于白色的空间内,丰富了空间结构的弹性,视觉上感受到空间立体的真实存在,映入眼帘的是古老而经典的色彩,让人觉得宁静、惬意,在白色玻璃灯光下映射出时尚的光泽,古典和现代气息形成的强烈反差,错综交杂,却又不失空间协调感。

Hong Kong Design Co., Ltd.

Antique white material in the space shuttle, the rich structure of flexible space, the visual feel of the real three-dimensional space, the eye is ancient and classic colors, make people feel calm and relaxed in a white glass Lights out stylish gloss mapping, the formation of classical and modern sharp contrast, the wrong cases are mixed, but without losing the sense of space coordinate.

地点:厦门市嘉禾路386号东方财富广场B栋403
建筑面积:230m²
室内设计:汤建松
摄影:刘腾飞

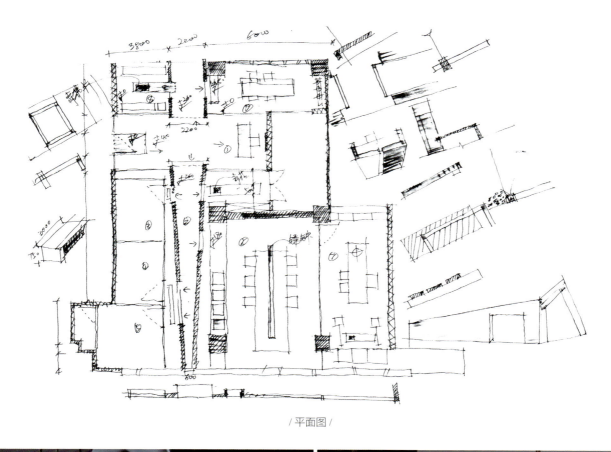

/ 平面图 /

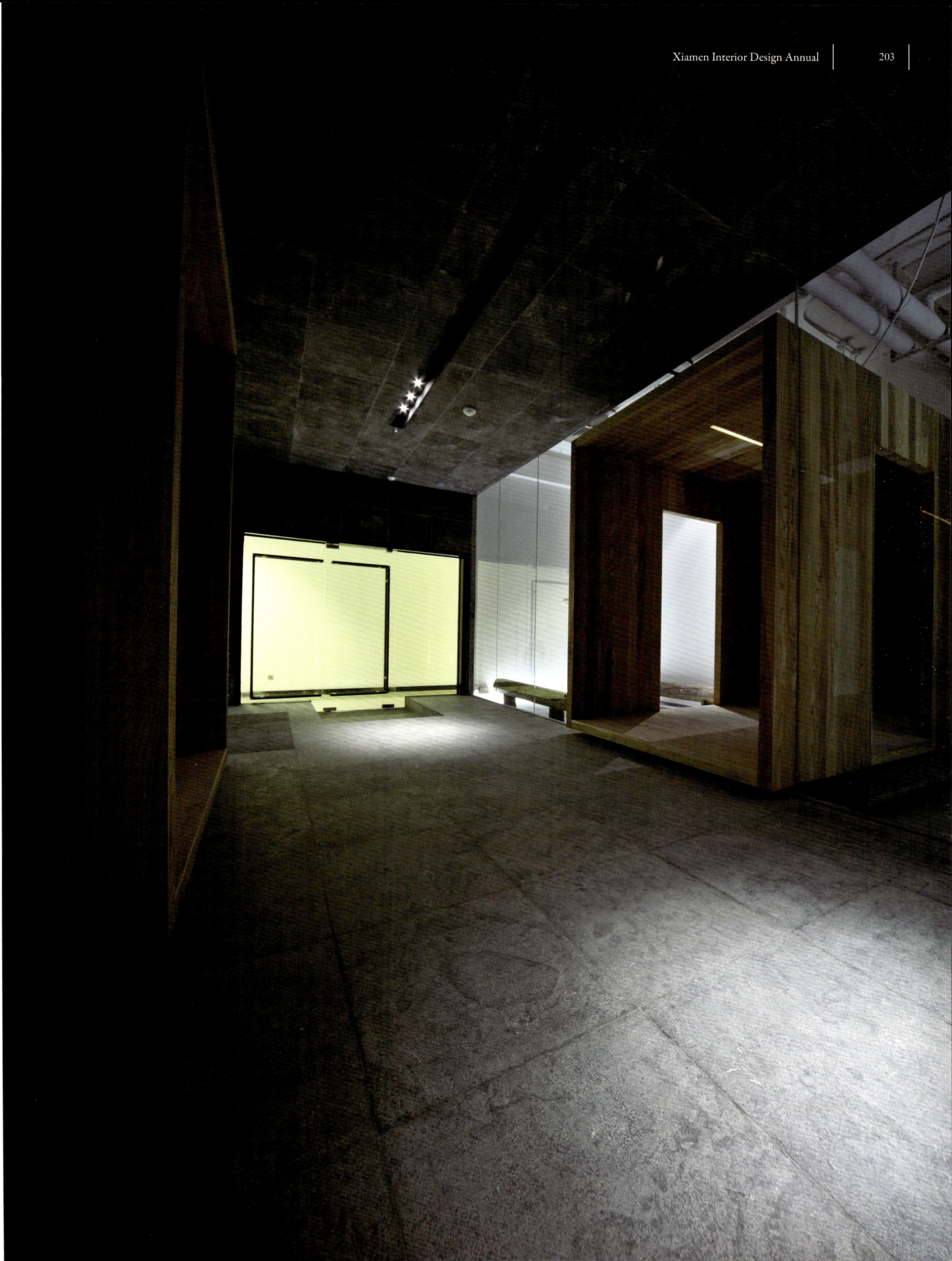

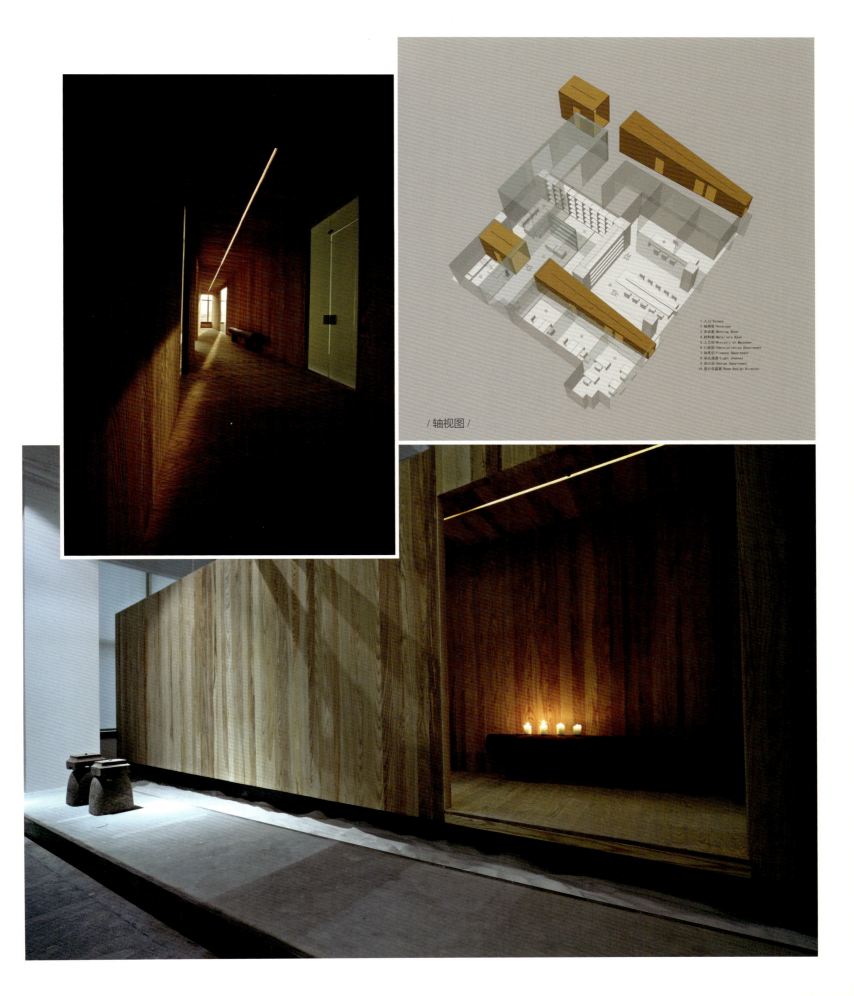

/ 轴视图 /

吴伟宏 Wu Weihong

- 宏盟东方室内设计机构_创意总监
- *Oriental interior design agency Omnicom_ Creative director*

个人经历

- 宏盟东方室内设计机构_创意总监
- 意大利米兰理工大学室内设计_管理硕士
- IAI亚太建筑师与室内设计师联盟_资深会员
- 中国建筑装饰协会室内设计分会_会员
- 中国建筑装饰协会_室内建筑师
- 福建省建筑装饰行业_优秀青年设计师

主要荣誉

- 2010年，亚太设计双年大奖赛提名；
  亚太设计双年大奖赛最佳酒吧空间设计大奖提名
- 2009年，荣获"大天杯"福建省室内与环境设计大奖赛公建类二等奖；
  荣获亚太室内设计大奖赛入围奖；亚太设计"中国风"大奖赛优秀奖
- 2008年，荣获"星辉杯"福建省室内与环境设计大奖赛公建类二等奖
- 2007年，荣获"辉煌杯"福建省室内与环境设计大奖赛公建类一等奖

Experience

- Oriental interior design agency Omnicom creative director
  Polytechnic University of Milan, Italy Master of Interior Design
  IAI Union Pacific Architects and interior designers, senior members
  Interior Design China Building Decoration Association Chapter Member
  China Building Decoration Association of Interior Architects
  Fujian outstanding young designers building decoration industry

Main Honours

- In 2010, the Asia-Pacific Design Biennial Grand Prix nominated;
  Design Biennial Asia-Pacific Space Design Competition Award nomination Best Bar
  In 2009, won the "Big Day Cup " Fujian Grand Prix Interior and Environmental Design Award category of public construction;
  won the Asia Pacific Interior Design Competition Finalist;
  Asia Pacific design "Chinese style " Grand Prix Award of Excellence
  In 2008, won the "Star Cup " Fujian Grand Prix Interior and Environmental Design Award category of public construction
  In 2007, won the "Cup glory"Fujian Interior and Environmental Design Competition first prize public buildings

每当看见一个作品诞生，重视别人的评价，无论是好还是坏，都是一种欣慰。这些观点，也是别人对我们设计的一种理解。这就是一种动力，一种设计的源泉。我们每次接到一个好的或者是不好的案子，我们都认真对待，认真对待每一件事情，希望作品尽早诞生。

贵司主要从事哪方面的设计？最新的作品是什么？

- 目前公司主要设计的项目有：夜店、会所、娱乐、餐饮类等。其中，夜店设计的是比较商业、比较市场的东西，这种空间的设计受甲方影响很大，因为甲方主要关注的是成本的回收，如果我们从设计的角度看问题的话，会和市场脱节。最近主要在设计几个餐厅的项目。

您的设计灵感来源于什么？

- 很多设计灵感来源于生活，是生活中点点滴滴的积累。比如：在旅游过程中，不经意见一些灵感的激发。我可以把几种不同生活方式的灵感结合，比如：设计一个夜店，我可以把它设计成一个比较生活化的餐厅；也会把一个餐厅设计成一个T型舞台或仓库；就是说，不一定业主要设计什么，我就把特定的环境设计成这个样子。

如何获得设计灵感？

- 业余时间多听大师讲座，出去走动，获得灵感。

设计思考枯竭的时候，怎么办？

- 在设计"瓶颈"时，一定放下手中的事情，完全放松自己。在有效的设计时间内，合理分配时间。

设计项目的内容和对未来的思考？

- 所设计的项目多集中在南方，未来计划走出去。目前，已经有部分北方的项目找过来了。早几年，就走出去。做设计主要靠灵感和喜欢，如果缺乏了设计激情，就很难从事好这个行业了。

您从事设计工作多久了？您的设计历程是什么？

- 快20年了。我曾经有一段时间去做工程，但始终没有考虑放弃去做设计，以前也开一些酒吧。
- 最初学习美术，以前做家具设计，后来一个偶然的机会，和我一个师傅（他是开设计公司的）一起帮朋友设计住宅，我起初是个帮手，后来，变成主笔做室内设计。真是太偶然了，让自己兴趣爱好发生转移。

- 目前，很多对我信任的业主都已经不要我做设计效果图。因为他们知道我设计的东西，而且现在效果图都比较假象，他们更多的是追求最后的结果。我和业主共同期待最终效果。我们最多就画一些草图。

设计感言？

- 每当看见一个作品诞生，重视别人的评价，无论是好还是坏，都是一种欣慰。这些观点，也是别人对我们设计的一种理解。这就是一种动力，一种设计的源泉。我们每次接到一个好的或者是不好的案子，我们都认真对待，认真对待每一件事情，希望作品尽早诞生。

您喜欢的设计师是谁？

- 喜欢菲利普·斯达克，从十多年前看到他的东西就非常东西，以前在一些杂志上看见，现在仍然喜欢。

BAG办公

生态办公设计，随着国家对环保的重视，室内设计师也寻找各种方法使工作环境更环保。透明度，敏捷性和可持续性的新的办公空间的一部分。办公空间应该拥有最高的性能和舒适性。家具，灯饰，油漆颜色和其他元素，来表现空间的个性。

BAG office

Bio-office design. Designer is looking for various ways to produce a environment-friendly office . The office should be comfortable and of various functions. And to show the personality by through the furniture, lighting, paint color and other elements .

项目地点：福建厦门软件园
项目面积：3500 m²
公司名称：厦门宏盟东方室内设计机构

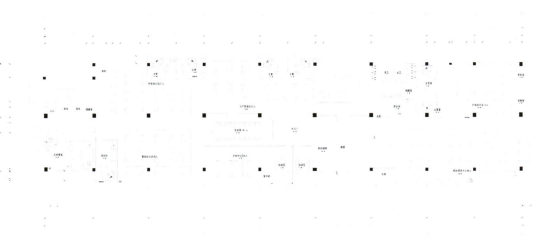

/ 四楼平面图 /

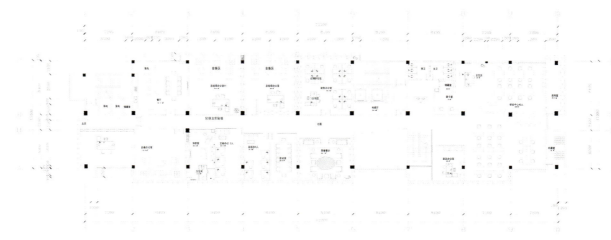

/ 五楼平面图 /

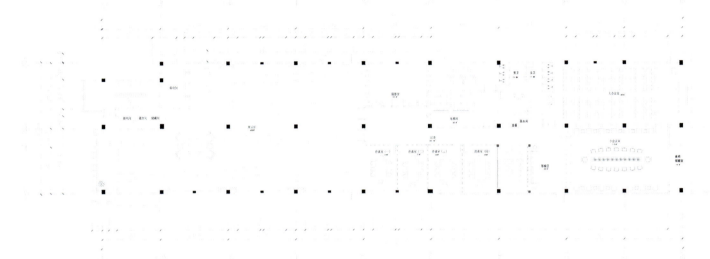

/ 六楼平面图 /

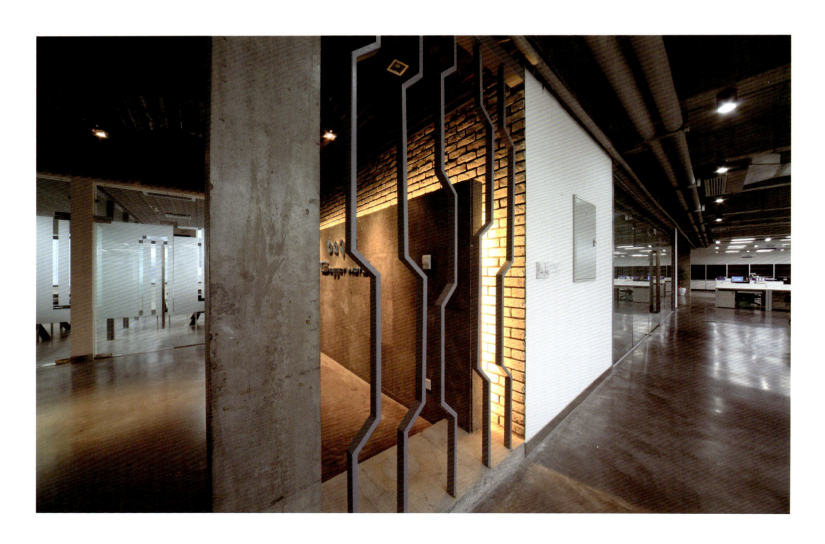

Bellagio餐厅

本案大型浅浮雕"波浪墙"的背景非常引人注目,它以一种安静的方式表现出漂移和游动,奇特舞台灯光让这整个空间充满梦幻的感觉,T台服务通道,就像一个走秀的舞台,不经意间体现设计师的不拘一格,独具匠心的设计理念。设计注重为使用者提供非正式交流空间,便利的休息设施充分体现了"以人为本"的理念。

Bellagio Restaurant

This huge relieve is attractive by it 'Wave Wall' background. The shows the drifting and swimming in a quiet way. The fancy arena lighting makes the space full of dreamy feeling. T shape service channel is just like a arena, shows the unlimited, spectacular designing idea unconsciously. The designer provide a informal communication space for users. The convenient leisure equipment shows the designing idea of 'People-Oriented'.

项目面积:400m²
公司名称:厦门宏盟东方室内设计机构
设计师:吴伟宏
摄影师:申强

代表作品(二)

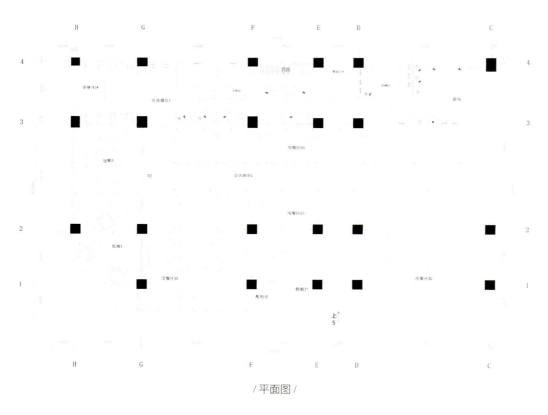

/ 平面图 /

Show吧

本案以"黑白"为主题，整个室内空间强调整体的纯净和利落。运用现代家具元素，表达了设计者要求现代时尚的意图，整个空间以几何面结构和线来组合，在墙面添加了多媒体。整个空间以网络的形式连接起来，使空间形成一个整体，周围到处都是投影与灯光营造一个活泼热情的气氛。钢板烤漆地板是暗灰色，表达设计者追求空间逆动的意境。

Show Bar

The design focuses on Black&White, it shows a clean and sharp environment. Designer makes it fashionable by using modern furniture. The whole space is of geometric plane and lines the designer connect the whole space by network, integrate the space together. With the projection and lighting, it creates a vivid, warm atmosphere. The baking finish iron plate floor is black gray, shows the designer's pursuit of space retroaction.

项目面积：450m²
公司名称：厦门宏盟东方
室内设计机构
设计师：吴伟宏
摄影师：申强

代表作品（三）

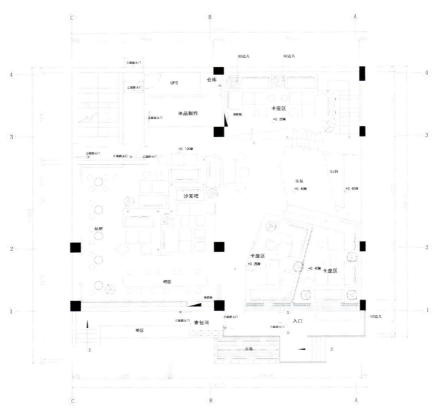

/ 一楼平面图 /

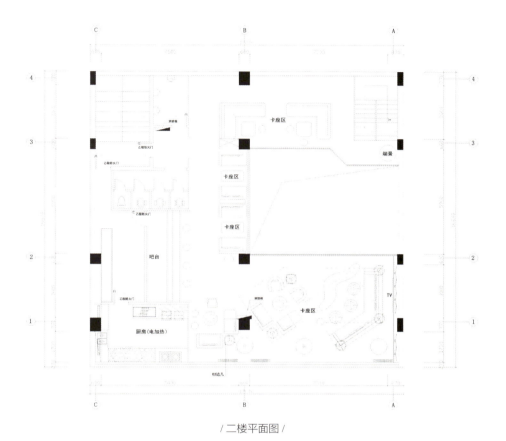

/ 二楼平面图 /

徐福民 Xu Fumin

- 厦门徐福民室内设计有限公司_设计首脑
- Xiamen Xu Fumin Interior Design Co., Ltd._ Creative director

个人经历

- 厦门徐福民室内设计有限公司_设计首脑
  厦门美筑设计工程有限公司_设计顾问
  毕业于福建师范大学艺术学院_装璜与装饰设计专业
  就读于厦门大学艺术学院_环境艺术设计专业艺术硕士
  中国建筑学会室内设计分会_会员
  中国室内装饰协会_室内设计师

主要荣誉

- 2010年获 中国室内大奖赛/商业工程类_二等奖
  2009年获 "金外滩"设计大奖赛/最佳办公空间（优秀）奖
  2008年获 中国室内设计大奖赛/办公工程类_三等奖
  2007年获 中国室内设计大奖赛/商业工程类_一等奖
                                        /办公方案类_一等奖
  2006年获 中国室内设计大奖赛/商业方案类_佳作奖
  2006年获 海峡两岸四地室内设计大奖赛/商业方案类_二等奖
  2005年获 海峡两岸四地室内设计大奖赛/住宅工程类_二等奖
  2004年获 中国室内设计大奖赛/住宅方案类_二等奖
  2004年获 福建省室内设计大奖赛/住宅工程类_优秀奖

Experience

- Xiamen Xu Fumin Interior Design Co., Ltd.　Creative director
  Xiamen Meizhu Design Engineering Co., Ltd.　Design Consultant
  He graduated from Fujian Normal University College of Art Design and decoration decoration
  Studied at the Art Institute of Xiamen University, Environmental Art Design Master of Fine Arts
  China Institute of Interior Design Branch Member
  Interior Design China Interior Decoration Association

Honor

- 2010 the Chinese Grand Prix indoor commercial engineering Second Prize
  2009 by the "Bund" best office space design competition (excellence) Award
  2008 China Interior Design Competition by the Third Class office project
  2007 by China Business Engineering Interior Design Competition First Prize
                            First prize of office programs
  2006 Business solutions in China Interior Design Competition Award of Eminence
  2006 Four cross-strait business solutions type Interior Design Competition Award
  2005 Cross-Strait Four Residential Engineering Interior Design Competition Award
  2004 China Interior Design Competition Award category of residential program
  2004 Residential Interior Design Competition in Fujian Province Engineering Excellence Award

如果暂时不做设计，自己最想去欧洲学习，感受生活。在理论的学习上，重新学习"建筑史"的相关课程，从理论上再提高自己，努力把空间的概念做到室内设计中。

从事设计工作多久了？

- 我04年到厦门开公司的。之前我在福州，96年开始从事设计工作，已经有十几年了。不过之前还做工程。

如何在繁杂的事务性的工作中，提高工作效率？

- ①设计管理；②多方面培养公司中其他的设计师；③设计公司不必要做的太大。
- 我现在很少加班了，体力损耗很大。我之前也经常加班到凌晨三四点，早上八九点又到公司。而白天大部分处理的都是杂事，日程安排缺乏管理。用于设计的精力就被分散了（设计师一定要先会生活，这样才能保持比较好的创作状态）。所以一定要管理好自己，安排好自己，减少不必要的工作，保证精力做设计，做到良性循环。

在设计过程中，如何与业主进行设计交流？

- 在设计过程中，很多设计的时间非常赶。就拿这个"地产类"的设计来说，这个项目施工四十多天就完成了，都是赶工的。业主他有自己的一个营销的计划，我们也比较无奈。本来我们都选好了软装，但是业主把之前样板房的家具和装饰物都拿过来，随意拼凑。可以说，我们原本的设计都被业主改掉了，理由很简单——节省资金。

如果不做设计，自己最想做什么？

- 如果暂时不做设计，自己最想去欧洲学习，感受生活。在理论的学习上，重新学习"建筑史"的相关课程，从理论上再提高自己，努力把空间的概念做到室内设计中。

你欣赏的建筑师、设计师是谁呢？对你的影响比较大？

- 路易斯·康；盖里；密斯·凡·德罗。我在厦大念艺术硕士的时候，导师对西方艺术史研究比较深，所以我也非常欣赏欧洲的大师们。

您的爱好是什么？

- 我基本上都没有爱好了。因为工作太累，白天下班后就去洗洗脚，放松一下，又回到公司做设计。因为晚上终于安静了，可以有一点灵感。

UTOP优伯科技办公楼

整体设计前卫，大胆，张扬。强调了视觉的冲击感。

办公楼共有两个楼面，一楼作为公共区域和各职能办公部门，会议室等。用几何体造型创造出隔而不离的各职能办公区，合理利用每一块空间的同时消隐了为数不少的房柱。二楼是加建的钢结构建筑，作为总经理的办公室和商务休闲区及大露台。在楼面上开口打通的位置，两个多边几何体在空间的相互穿插，共同创造出最具特点和独创的楼梯造型。

Technology office UTOP

Overall design of the avant-garde, bold and assertive. Emphasizes the visual impression.

There are two floor office building, first floor functions as a public area and all offices, a conference room. Modeling using geometry to create a compartment and separated from the various functions of the office, the rational use of space, while each piece blanking of the large number of room and pillar. On the second floor is the addition of steel construction, as the general manager's office and the business and leisure areas and large terraces. Openings in the floor opened up the position of the surface, the two multilateral geometry in space, interspersed with each other and jointly create the most characteristic and original staircase shape.

主创设计：徐福民
设计公司：厦门徐福民室内设计有限公司
摄影：吴永长

代表作品（一）

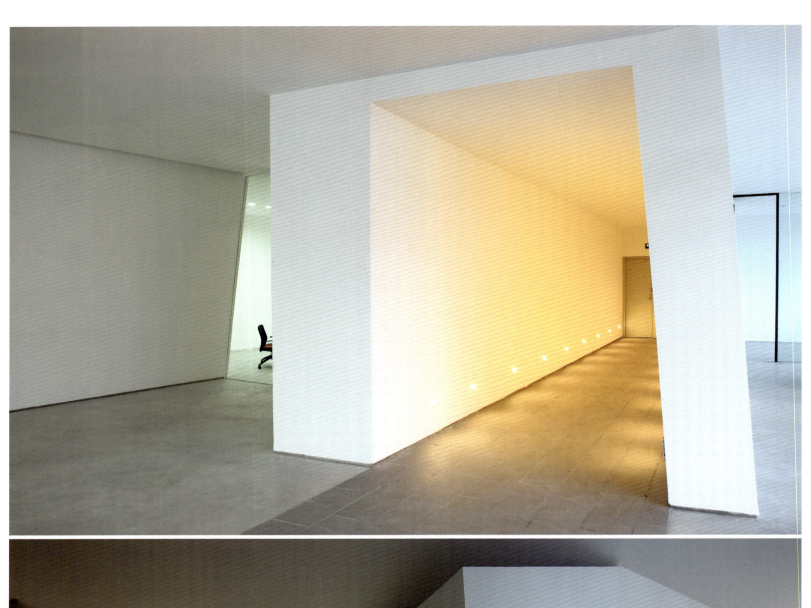

通过钢结构骨架,两个多边几何体和屋面的天窗有机连成一体,蓝天白云阳光自上洒下……

以公司的文化理念为设计的基础,造型,功能和结构协调统一。利用灰色的主调,白色的几何体墙面,透明玻璃打造了一个极富现代感的工作空间。

To the corporate culture as the basis for the design, shape, function and structure of harmonization. The main theme of the gray, white wall geometry, transparent glass to create a very modern work space.

Through the steel structure, the two multilateral geometry and roof skylight organic fused. Tear down the sun from the sky … …

中骏·财富中心售楼处

房产销售中心的空间设计以功能性为主要目的。如何从形态和空间上脱离传统销售中心的空间模式，是设计师在本项目中致力突破的环节。本中心的设计颠覆了单一空间概念，在合理分配使用功能的基础上，将"水"这一设计元素延伸性地融入空间的塑造，营造出未来与现实并存的超现实魔幻意境。通过"水"，来表达祥和富贵之意以及人与自然和谐共处的绿色人居概念。

Fortune Plaza Sales Center

Real estate sales center space designed to function as the main purpose. How to shape and space center space away from the traditional sales model, a designer in this project to break the link. The Center's design to subvert the concept of a single space, use a reasonable allocation of functions, based on the "water" extension of the design elements integrate into the shaping of space, creating a future of coexistence with the surreal reality of magical mood. Through the "water " to express the meaning of wealth and harmony between man and nature live in harmony concept of green living.

主创设计：徐福民 陈朝晖
设计公司：厦门美筑设计工程有限公司
摄影：吴永长

Xu Fumin 代表作品（二）

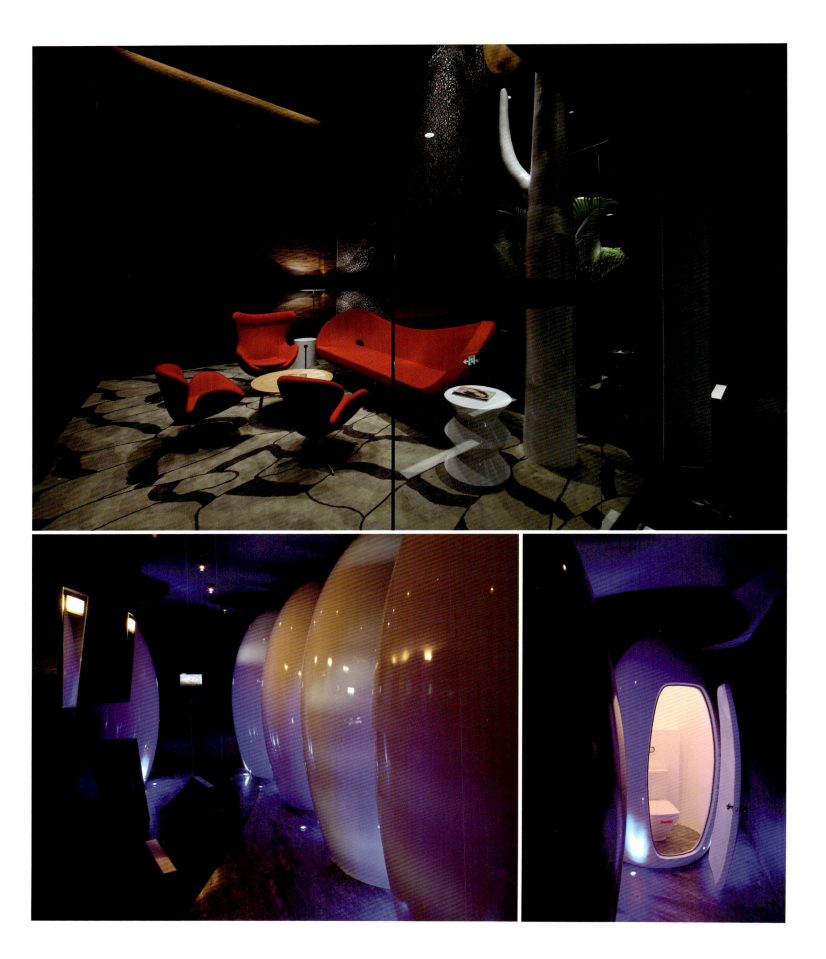

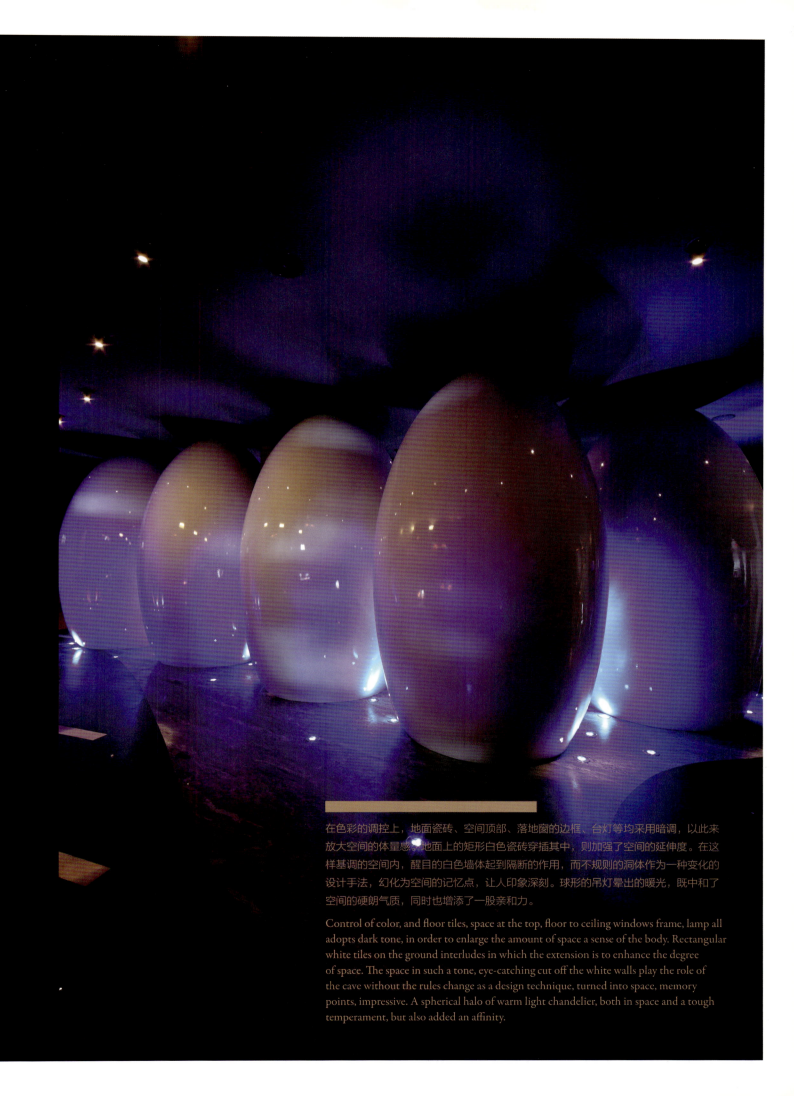

在色彩的调控上,地面瓷砖、空间顶部、落地窗的边框、台灯等均采用暗调,以此来放大空间的体量感。地面上的矩形白色瓷砖穿插其中,则加强了空间的延伸度。在这样基调的空间内,醒目的白色墙体起到隔断的作用,而不规则的洞体作为一种变化的设计手法,幻化为空间的记忆点,让人印象深刻。球形的吊灯晕出的暖光,既中和了空间的硬朗气质,同时也增添了一股亲和力。

Control of color, and floor tiles, space at the top, floor to ceiling windows frame, lamp all adopts dark tone, in order to enlarge the amount of space a sense of the body. Rectangular white tiles on the ground interludes in which the extension is to enhance the degree of space. The space in such a tone, eye-catching cut off the white walls play the role of the cave without the rules change as a design technique, turned into space, memory points, impressive. A spherical halo of warm light chandelier, both in space and a tough temperament, but also added an affinity.

个人经历

- 注册高级室内建筑师
  U空间室内设计师国际联盟_副主席
  厦门东方设计装修工程有限公司_设计总监
  厦门东方设计研究院_副院长

获奖情况

- 2010年中国国际空间环境艺术设计大赛酒店空间工程类优秀奖
  2010年度金堂奖十佳酒店空间设计作品
  2010亚太室内设计双年大奖最佳酒店空间设计大奖提名
  2010亚太室内设计双年大奖最佳酒吧空间设计大奖提名
  2002年第二届欧典杯全国居室装饰实例大赛获佳作奖
  2001年吉事多中国首届卫浴空间设计大赛户获优秀奖

主要作品

- 厦门U空间精品酒店
  厦门皇家艾美酒店（深化设计）
  厦门福隆体育公园（国家乒乓球队南方训练基地）
  福州西湖大酒店（行政咖啡厅、接见厅、总统套房改造）
  福建人民大会堂（地区议事厅）
  武夷山茶苑大酒店
  厦门尚帅汽车旅馆
  厦门东方巴黎
  东方明珠
  东方丽景售楼部样板房
  厦门巴厘香泉度假公寓样板房及厦门巴厘香墅

Experience

- Registered Senior Interior Designer
  U space interior designer, the League of Vice-Chairman
  Xiamen Dongfang design and decoration Engineering Co., Ltd._Creative Director
  Vice President of Design and Research Institute of Xiamen Oriental

Awards

- 2010 China International Environmental Art Space Hotel Space Engineering Design Competition Award of Excellence
  Top Ten Hotels Award 2010 Jintang work space design
  2010 Asia Pacific Interior Design Biennial Awards Award Nomination Best Hotel Interior Design
  2010 Asia Pacific Interior Design Biennial Awards Award Nomination Best Bar Interior Design
  EU Code of 2002, the second instance of home decoration contest Cup won Honorable Mention
  Jishi Duo 2001, China's first indoor bathroom space design competition won an Outstanding Award

Honor

- Room boutique hotel in Xiamen U
  Xiamen Royal Meridien Hotel (depth design)
  Xiamen Fulong Sports Park (national table tennis team training base in the south)
  Fuzhou Lakeside Hotel (administrative coffee shop, meeting rooms, Presidential Suite Renovation)
  Fujian Great Hall of the (region Chamber)
  Gametea Hotel Wuyishan
  Xiamen is still handsome Motel
  Xiamen Oriental Paris
  Pearl of the Orient
  Department of Housing sales model for the East Belleview
  Bali xiangquan holiday apartment in Xiamen and Xiamen model room villa in Bali, Hong

> 厦门是一个有历史积淀的城市，并具备包容的文化底蕴，厦门设计应该走出去，中国设计亦然。

贵司主要从事哪方面的设计？最新的作品是什么？

- 主要从事豪宅、酒店、会所。最近作品U空间精品酒店，巴厘香泉度假公寓样板房…

您的设计灵感来源于什么？

- 大自然与建筑。

您从事设计工作多久了？您的设计理念是什么？

- 我从事设计工作20多年了。我的设计理念是充分满足功能的同时最大化的将自然融入空间。

在设计过程中，您最注重哪方面的工作？对于各种设计元素有没有什么偏好？

- 我最注重空间及功能的整合，因为空间是一个好的室内设计的灵魂。而对于表面的装饰元素，只要对案例合适我就会用。

从前期接触项目、出概念、做方案到完成整个项目，觉得最大的困难是什么？

- 比较新的创意不容易被接受。

欣赏的或者对您影响最大的人是谁？或是某种风格、思潮、理念？

- 贝聿铭与安藤忠雄。推崇建筑融合自然的空间理念。

怎么看待厦门设计？以及中国设计？

- 厦门是一个有历史积淀的城市，并具备包容的文化底蕴，厦门设计应该走出去，中国设计亦然。

未来的公司会如何发展？您本人呢？

- 坚持经营方向的定位，把每个案例当作品做。我个人则要保持身心的健康，与时俱进，活到老，学到老，做到老。

厦门U空间精品酒店

这是一个只有10个房间的小精品酒店，座落于厦门顶级别墅区"巴厘香墅"。是U空间室内设计师国际联盟为方便设计师交流而投资的小酒店，除了10个房间，还附设大堂吧、小咖啡厅、小会议室。因为是面对设计师群体的空间，设计师大胆选用了灰色做为主调，设计干净利落、酷感十足，没有多余的装饰，只有雕塑家龚栋的群雕"我们的天空"散置于空间中，引人无限联想。

U space for boutique hotel

This is one of only 10 rooms in the small boutique hotel, is the International Union U space interior designers and investment to facilitate the exchange of small designer hotels, in addition to 10 rooms, is also attached to the lobby bar, coffee shop, small conference rooms. Because space in the face of designer groups, designers chose a bold gray as the main theme, the design neat, cool feeling full, no frills, only the sculptor's statues Dong Gong, "the sky" casual place Space, the introduction of the unlimited.

主创设计：曾冠伟
设计公司：厦门东方设计装修工程有限公司

代表作品（一）

大堂、酒吧、咖啡厅采用全开放的布局，利于设计师举办活动时的互动交流。用U空间联盟的"U"字作为造型元素来区隔功能区域。一个悬浮的侧倒的"U"形吧台区分了大堂接待区与大堂吧，一个平面的"U"形酒水柜区分了大堂吧与小咖啡厅，满足了功能需求又形成了视觉焦点，既隔又透，富有设计感。

Lobby, bar, coffee shop layout with wide open, which will help designers of interactive communication when organizing activities. U space alliance with the "U" word as design elements to separate functional areas, a suspension on its side of the "U" shaped bar for a distinction between the lobby and lobby bar, then holding area, a flat "U"-shaped drinks cabinets distinction between Lobby bar and coffee shop to meet the functional requirements and form a visual focal point, both across and through, full of design sense.

整个酒店的用材简单又自然，瓷砖、德国环保型复合地板、西班牙灰泥、手工漆、白蜡木，所有材料都统一在灰色调里，和谐又现代。偶尔出现的红色坐椅、灯具、洗手间明黄的玻璃隔断及窗外的绿树，给空间带来了跳跃的活泼感。

The hotel is simple and natural timber, ceramic tiles, the German environment-friendly flooring, Spanish plaster, hand paint, ash wood, all materials are uniform in gray with a harmonious and modern. Occasional red chairs, lamps, bathroom bright yellow in the glass partition and trees outside the window, jump to space has brought the lively feeling.

巴厘香墅江府

"黑白灰"的建筑外观,很"巴厘"也很现代。设计师决定将这一色调延伸进室内,以爵士白云石满铺地面,配以白墙清玻为主调,纯静而优雅;局部黑白根云石及黑色烤漆玻璃,黑色开放漆面的对比,因为选材的精致玩出了黑白调永不过时的精彩;白橡木的自然本色给空间带来了家的温馨感觉;而红色沙发及休闲椅的点缀,欧式新古典不锈钢吊灯及墙上美洲画家的人体油画与中国的漆画都尽显当下流行的混搭之风。通过对材料、家具、灯光、配饰的巧妙搭配,设计师创造出一个低调奢华的现代空间。

Bali Villas

"Black and white ash, "the building exterior, it is the "Bali"is also very modern. Designers decided to extend the color into the room to shop floor full of jazz dolomite, together with the white walls and clear glass as the keynote, critical system and elegant; local black Baigen Yun Shi and black painted glass, black paint in the open Contrast, because the selection of the fine play out of the wonderful black and white never go out of tune; white oak natural qualities brought to the room warm feeling of home; and the red sofa and lounge chair the decorations, neo-classical European-style chandeliers and wall stainless steel American painter of the human body oil painting and Chinese painting are filling mix and match current popular style. On materials, furniture, lighting, accessories ingenious, low-key luxury designer to create a modern space.

项目地址:曾冠伟
设计单位:厦门东方设计装修工程有限公司

代表作品(二)

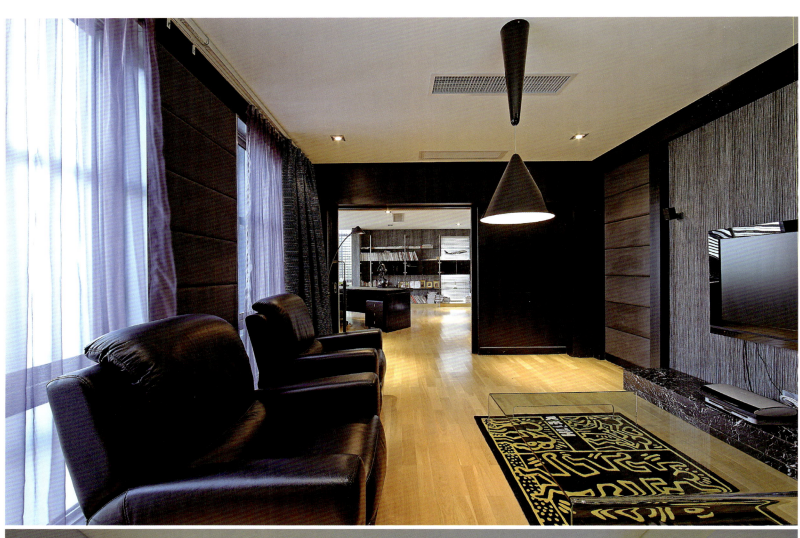

别墅整体通透与明亮,室内外融为一体。设计师刻意模糊室内外的空间界限,在首层公共空间中将传统空间所定义的"墙"拆除,并将为保证运动量而设计的细长泳池,紧靠餐厅设置,让客厅、餐厅、厨房、吧区、泳池边的休闲阳台在视觉中没有界限,互动融洽而有趣味。自然的光线从室外倾泻进来,丰富了空间层次的光影变化,让人感受到空间的延展性及无拘无束的休闲氛围,于简洁中体验终极的舒适优雅。

Villa overall transparent and bright, indoor and outdoor integration. Designers deliberately blurred boundaries of indoor and outdoor space, public space in the ground floor space will be defined by the traditional "wall" removed, and designed to ensure the exercise slender swimming pool, close to the restaurant setting, so the living room, dining room, kitchen , Bar area, the leisure pool there is no balcony in the visual boundaries. Light coming from the outside, rich light level changes in the space, people feel the space, scalability and unfettered leisure atmosphere, experience the ultimate in simple and elegant comfort.

郑传露 \ Zheng Chuanlu

- 厦门市共想装饰设计工程有限公司_创意总监
- Xiamen Gongxiang Decoration Engineering Co., Ltd._Creative Director

郑传露
- 1976年出生于福建厦门，1998年毕业于厦门理工大学。
- 2001年，取得助理工程师任职资格；
- 2003年，取得助理工艺美术师任职资格；
- 2003年，取得中国室内设计师资格认证。
- 2001年—2005年，任职于厦门喜玛拉雅装饰设计工程有限公司；
- 2006年，成立厦门市共想装饰设计工程有限公司。
- 2006年，14届APIDA亚太室内设计大奖商业组铜奖；
- 2007年，15届APIDA亚太室内设计大奖办公组荣誉奖；
- 2009年，获得17届APIDA亚太室内设计大奖赛商业组银奖；
- 2010年，获得18届APIDA亚太室内设计大奖赛办公室组银奖。

Zheng Chuanlu
- Born in 1976, Xiamen, Fujian, graduated from Xiamen University in 1998.
- In 2001, Assistant Engineer qualifications achieved;
- In 2003, the qualification obtained assistant craft artists;
- In 2003, China's interior designers to obtain certification.
- In 2001 -2005, working in the Himalayan Decoration Engineering Co., Ltd. Xiamen;
- In 2006, the establishment of Xiamen Gongxiang Decoration Engineering Co., Ltd..
- In 2006, the 14th Asia Pacific Interior Design Awards APIDA business Bronze Award;
- In 2007, the 15th Asia Pacific Interior Design Awards office APIDA Group Honor Award;
- In 2009, received the 17th Asia Pacific Interior Design Competition APIDA Silver Business Group;
- In 2010, received the 18th Asia Pacific Interior Design Competition Office APIDA group Silver.

朱鹭欣 \ Zhu luxin

- 厦门市共想装饰设计工程有限公司_创意设计
- Xiamen Gongxiang Decoration Engineering Co., Ltd._Creative Design

朱鹭欣
- 2006年至今，任职于厦门市共想装饰设计工程有限公司。
- 2009年，获得17届APIDA亚太室内设计大奖赛商业组银奖；
- 2010年，获得18届APIDA亚太室内设计大奖赛办公室组银奖

Zhu luxin
- Since 2006, a total of like working in Xiamen Gongxiang Decoration Engineering Co., Ltd..
- In 2009, received the 17th Asia Pacific Interior Design Competition APIDA Silver Business Group;
- In 2010, received the 18th Asia Pacific Interior Design Competition Office APIDA group Silver.

庄钰璘 \ Zhuang Yulin

- 厦门市共想装饰设计工程有限公司_创意设计
- Xiamen Gongxiang Decoration Engineering Co., Ltd._Creative Design

庄钰璘
- 2002-2007年，任职于厦门喜马拉雅装饰设计工程有限公司；
- 2008年至今，任职于厦门市共想装饰设计工程有限公司。
- 2009年，获得17届APIDA亚太室内设计大奖赛商业组银奖；
- 2010年，获得18届APIDA亚太室内设计大奖赛办公室组银奖

Zhuang Yulin
- 2002-2007, working in the Himalayan Decoration Engineering Co., Ltd. of Xiamen;
- Since 2008, a total of like working in Xiamen Gongxiang Decoration Engineering Co., Ltd..
- In 2009, received the 17th Asia Pacific Interior Design Competition APIDA Silver Business Group;
- In 2010, received the 18th Asia Pacific Interior Design Competition Office APIDA group Silver.

厦门的市场较小，并不成熟，好的设计作品不易实现。每年我自己比较满意的项目就只有一两个。厦门业界的交流不多，没有形成切磋和互相学习的氛围。我自身也很少跟业界沟通，我们公司也很少自我包装、推广。为了接触更多的项目，我们也在慢慢地改善，比如参加比赛。这也是对自己的交代，是对自己的肯定。

您从事设计工作多久了？贵司主要从事哪方面的设计？

- 1995年，我初涉这一行。1998年，大学毕业后真正开始从事这一行。我们主要做的是一些商业空间。之前做茶庄，后来做写字楼，这两年做商业空间、零售、办公空间比较多。最近，在设计一个奢侈品店。跟不同的业主接触，会学到很多东西，所以每个行业我都希望有所涉及。我们接触的项目类别比较广，但是项目数量上不是那么多。根据公司规模，我们也会有选择性的接项目。

欣赏的或者对您影响最大的人是谁？抑或某种风格、思潮、理念？

- 早期的时候，安藤忠雄对我的影响较大。以前做设计比较注重形式，现在我们关注多元性。除了建筑、室内设计资讯，我也很关注平面设计、工业设计。大量的阅读使我受益匪浅。

您的设计理念是什么？

- 以前我可能更注重设计的形式，追求设计感。但是这几年我的观念逐渐改变。现在我注重的是功能性和在空间的感觉。比如，我现在的新办公室，空间设计很简单，但是很舒服，使人愿意待在这里。其次，我不会固定自己的风格，做方案之前，我们会对业主、项目包括市场地位做很多了解和分析，然后选择最适合的风格。

在设计过程中，您最注重哪方面的工作？

- 我注重空间的层次，讲求一种建筑感。很多室内空间是装饰性的，但是我会先做空间的结构，我喜欢比较干净简单的空间，不会过多的修饰。

- 我比较关心软装方面。厦门没有很专业的软装公司，或者无法达到我们的要求，所以只能自己挑选搭配，非常辛苦。厦门的设计配套市场不健全，所以我会去上海、香港等地，了解资讯和市场。

如何保证项目的实现效果？

- 我喜欢设计，所以一直坚持这项工作。为了保证效果，我们也会兼做施工。因为施工单位是以利益为第一位，其次才是效果。细微的改动都会影响整体效果。

- 有时候，我对一些项目非常感兴趣，也有可能会把工程的利润补贴设计费，以求争取到这个项目。

- 其实项目的实现度，跟客户的信任也有很大关系。我会有意识的培养自己现有的客户，选择客户，更好地确保方案的实现度。因为喜欢而接项目，不纯粹追求商业利益。

怎样和业主沟通、合作？

- 跟业主的沟通及建立信任和合作并不是一件易事。

- 设计师应该是解决问题的人——帮助客户解决问题。多跟业主沟通，多做前期工作，对品牌和市场进行分析，了解品牌定位，市场需求，甚至涉及营销的工作；熟悉空间的功能性，监督设计效果的实现。用最省的方式达到客户所需求的东西，无论是费用、实用还是效果，达到他满意的效果。其次，回访很重要。回访的过程中可以知道自己的不足，检验自己的设计及概念。

未来的公司会如何发展？你们个人呢？

- 我的精力有限，我现在比较注重培养公司的设计师。现在80后的设计师都很不错，我很喜欢跟他们在一起。他们比较注重形式感，更有个性，有想法。

- 我很喜欢做设计，也很享受。我希望一年有几个代表性的作品，剩下50%左右是一些盈利性的项目。但是我要保证每个项目都认真去做。这几年我还是会继续做自己想做的事情，我家人很支持我。这个行业很辛苦，天赋很重要，勤奋更重要，一定要坚持。

怎么看待厦门设计？

- 厦门的市场较小，并不成熟，好的设计作品不易实现。每年我自己比较满意的项目就只有一两个。厦门业界的交流不多，没有形成切磋和互相学习的氛围。

- 我自身也很少跟业界沟通，我们公司也很少自我包装、推广。为了接触更多的项目，我们也在慢慢地改善，比如参加比赛。这也是对自己的交代，是对自己的肯定。

黎柏洋服

最自然的就是最奢华的！设计师认为从天然的材质和简约的色调中衍生出人与自然、环境相互适应，在自然"美"中提炼特质是最难做到的。

本案设计充分考虑到延续建筑本身异形几何结构进行规划，将柴火堆中拾来的干树枝、废旧的老墙砖等纯天然材料重新整合利用，光线与异形的几何形体配合，使得整个空间更具有节奏感，折射出"黎柏洋服"量体裁衣、个性化定制服装的主题，真正体现表里如一的特质。

Libo clothing store

The most natural is the most luxurious! Designers believe that from natural materials and colors derived from simple man and nature, the environment adapt to each other, in a natural "beauty" to extract characteristic is most difficult to achieve.

Continuation of the case design fully into account geometry of the building itself shaped plan, the wood pile to pick up the dry branches and other waste of the old wall to re-integrate the use of natural materials, light and shaped with the geometry, making the space more With a sense of rhythm, a reflection of "Li Bai tailor" tailored, personalized custom clothing theme, and really say what the character.

主创设计：郑传露
设计师：朱鹭欣
设计公司：厦门市共和装饰设计工程有限公司

代表作品（一）

258 | Xiamen Interior Design Annual

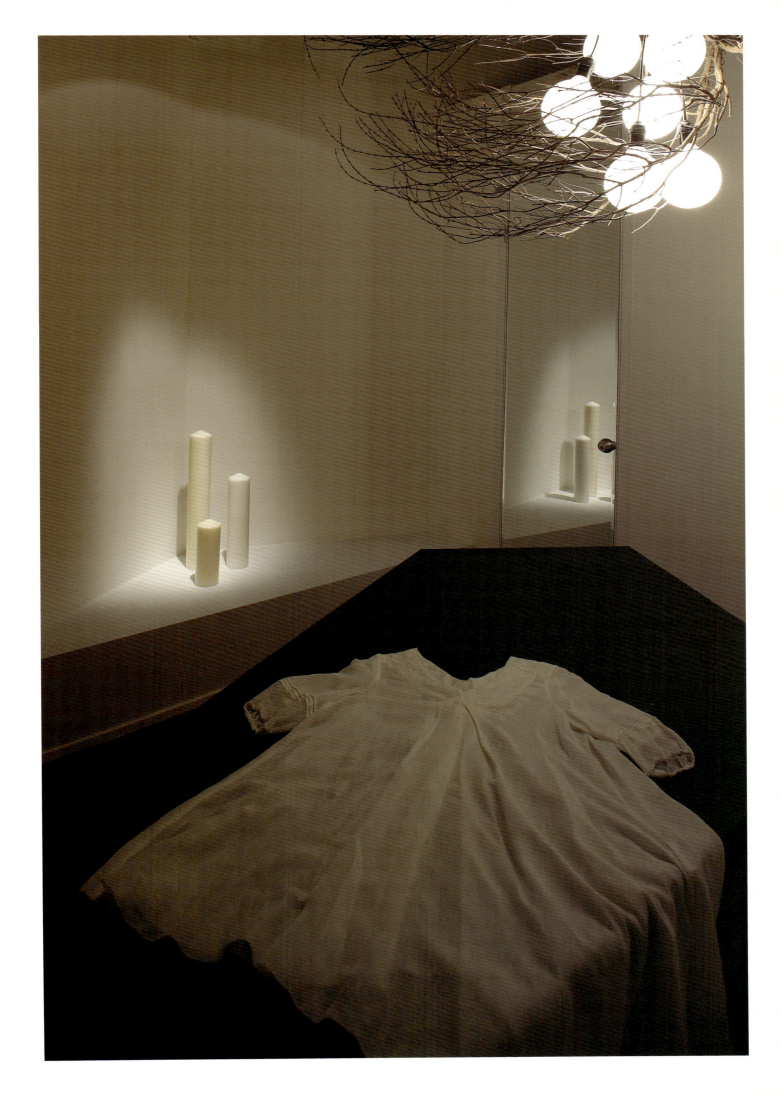

厦门共想装饰设计公司办公室

"作者"是一支充满激情,且热爱自然的设计团队,本次新办公室选址与"万国建筑博览"之岛的鼓浪屿隔海相望。本案开放的办公空间里,应用黑色的线条来强化空间的结构,废旧的木头、朴实的砖墙、蔚蓝的窗外,舒心且宁静!
都说设计来源于自然,来自于对生活的理解!本案中设计师带着这种特有的心态去感悟每一个思维的创造。

Gongxiang Decoration Design Co., Ltd.

"Author" is a passion and love for nature design team, this new office site and the "International Architecture Exhibition," the Gulangyu Island across the sea. Case open office space, application of black lines to strengthen the structure of space, waste wood, plain brick walls, blue windows, comfortable and quiet!

Said the design came from nature, from the understanding of life! Designer case with this unique state of mind to create the perception of each thought.

主创设计:郑传露
设计师:朱鹭欣
设计公司:厦门市共想装饰设计工程有限公司

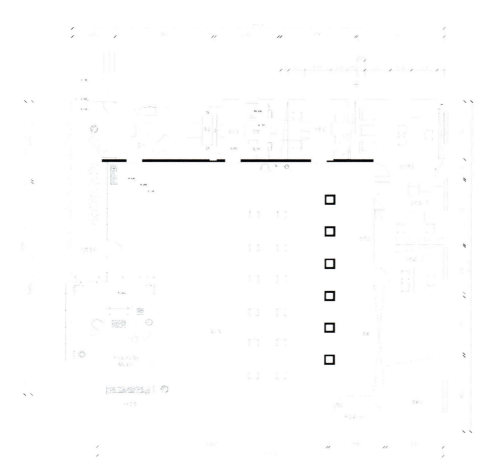

/ 平面图 /

厦门设计的觉醒

前些年，福州的设计师非常活跃，频频在国际国内的各种比赛中获奖，而同属福建的厦门却相对非常冷寂，少有全国知名的设计师。但近些年，特别是近两年来，厦门设计师开始觉醒，其活跃程度丝毫不亚于福州设计师。2010年，厦门的新生代设计师方令加、汤建松横扫香港亚太室内设计大奖，引起了设计界的普遍关注，他们也在一夜之间成为设计界的新星。这显示了香港亚太室内设计大奖强大的造星能力。其实，厦门设计师的获奖并不是开始于他们，早在多年前，厦门的设计师郑传露就连续几年获得了香港亚太室内设计大奖。

厦门不大但很精致，这是我印象中的厦门。鼓浪屿上散落了许多设计非常特别的殖民建筑，这些建筑在潜移默化中熏陶了厦门的设计师，使厦门的设计师不经意间获益良多。鼓浪屿的家庭旅馆非常出名，漫步在鼓浪屿，你会经常不经意间发现许多设计非常考究的家庭旅馆，这其中不乏出自厦门设计师的作品。如黄振耀就在鼓浪屿上自己投资并设计了PARK.酒店。厦门与台湾隔海相望，在厦门可以看到很多台湾的痕迹。厦门设计界也得益于地理位置的便利及文化上的相近性，与台湾设计界交流互动的比较早，也比较频繁。正如深圳设计师受香港设计师影响比较大，厦门设计师受台湾设计师的影响也非常大。也许是巧合，厦门设计师与台湾设计师都比较内敛，都默默耕耘，埋头做设计。这也许是前些年我们不太注意厦门设计师的原因之一吧。

2007年，在中国建筑学会室内设计分会的长沙年会上，我认识了从福州去厦门发展的设计师徐福民。这是我认识的第一个厦门设计师。徐福民说话不急不慢，光头，相当醒目。也许是因为与福州设计师走得比较近的原因，徐福民比较多的参加国际国内的各种评比比赛。早在多年前，就获得了多个国内大奖的金奖，是厦门设计界较早为业内所认识的设计师。

2009年，认识了厦门设计界活跃人物陈方晓，初见陈方晓觉得这人有点侠气，但也许正是这种江湖侠气使他能够为厦门设计界默默付出，使厦门设计师的凝聚力得到了极大的增强，也逐渐地让厦门设计界变得热闹，与外界有了更多的互动及交流。

提起厦门设计界，似乎不得不提孙建华、李学峰、孙少川。虽然由于种种原因，本书没有收录他们的作品，这是我们的一大遗憾。孙建华的公司是厦门设计界规模最大的公司之一，他不仅仅在厦门发展，更是在深圳设立了分公司，在全国的舞台上实践自己的设计梦想。孙少川是厦门设计界的奇才，没有经过专门的设计培训，通过自己的苦心钻研，在材料及技术方面有非常独到之处。据说，其在施工图方面的本领更是超然，他从来不去现场监督施工，因为其施工图的详细程度足以令现场施工无需担忧。李学峰我不熟悉，不敢妄加评论。

最后，我想谈谈厦门设计界的后起之秀，新生代设计师方令加、汤建松。方令加不太爱说话，比较喜欢收集散落在民间的物件，用在设计中，别有味道。汤建松相对活泼一些，比较关注当下现实，设计的抱负也比较远大，他近来的作品，大多概念性、展示性非常强。

杨琳、邵力中、姜辉是厦门设计界的逍遥派。他们学建筑出身，多年来，他们更多致力于公共建筑的设计，在这些实践中，需要承担的也许更多是社会责任。而设计理想的实现也许只有在一些小项目上，比如他们完成的"九间房七方院"。但他们更让我记住的，也许不是他们的项目，而是他们的生活态度。他们活得很优雅，设计也不急不慢，这才是理想的设计师生活。

厦门设计崛起之所以值得关注，是因为它不仅仅是几个设计师的浮出水面，而是一群人的集体亮相。这是一个时代的开始。可以毫不夸张地说，厦门设计是中国设计的一个缩影。

孔新民
2011年4月

图书在版编目（CIP）数据

2010厦门特辑（中国室内设计年鉴）：汉英对照 / 孔新民主编.
——北京：
中国林业出版社，2011.4
ISBN 978-7-5038-6134-5

Ⅰ.①2… Ⅱ.①孔… Ⅲ.①室内装饰设计—厦门市—2010—年鉴—汉、英 Ⅳ.①TU238-54

中国版本图书馆CIP数据核字(2011)第060379号

主　编：孔新民
策　划：纪　亮
编　写：孔新民　张　岩　王　超　刘　杰　孙　宇　李一茹
　　　　杨　丹　刘　嘉　张　雪　李　锐　祝　贺　高子涵
　　　　姜　琳　赵天一　李成伟　王琳琳　王为伟　李金斤
　　　　王明明　石　芳　王　博　徐　健　齐　碧　宋晓威
　　　　张文媛　陆　露　何海珍　刘　婕　夏　雪　王　娟
　　　　黄　丽　程艳平　高丽媚　汪三红　肖　聪　张雨来
　　　　陈书争　韩培培　付珊珊　张　雷　傅春元　邹艳明
　　　　高囡囡　杨微微　姚栋良　武　斌　陈　阳　张晓萌
　　　　魏明悦　佟　月　金　金　李琳琳　高寒丽　赵乃萍
　　　　裴明明　李　跃　金　楠　邵东梅　李　倩　左文超
　　　　陈　婧　陈圆圆　陈科深　吴宜泽　沈洪丹　韩秀夫
　　　　高晓欣　包玲利　郭海娇　阮秋艳　王　野　刘　洋
　　　　关丽楠　刘柏先　黄艳平　苏秀颖　唐　飞　胡　阳
　　　　牟婷婷　朱　博　宁　爽　刘　帅　梁　爽　曹　英

中国林业出版社 · 建筑与家居出版中心

出版联系：纪　亮　李　顺
联系电话：010-8322 3051
在线对话：1140437118（QQ）

出版：中国林业出版社
（100009 北京西城区德内大街刘海胡同 7 号）
网址：www.cfph.com.cn
E-mail：cfphz@public.bta.net.cn
电话：（010）8322 3051
发行：新华书店
印刷：恒美印务（广州）有限公司
版次：2011年4月第1版
印次：2011年4月第1次
开本：230mm×300mm
印张：17
字数：200千字
定价：260.00（RMB）；50.00（USD）